International Research on Poverty Reduction

Series Editors

Xiaolin Wang, Fudan University, Shanghai, China

Qianqian Liu, Finance Center for South-South Cooperation, Hong Kong, China

This book series covers poverty reduction practices and experience, knowledge mobilization, poverty reduction strategy analysis, and comparative research of different poverty reduction models. By analyzing China's poverty reduction approaches and experiences, and comparing them with those of other countries, this book series provides policy makers, researchers and practitioners in poverty reduction with theoretical and technical support, as well asproviding a new channel for sharing experiences as well as facilitating collaboration and communication in poverty reduction development fields between China and the rest of the world. Poverty remains a crucial development issue for the world, and China has become an important player in the global endeavor in poverty reduction and development. China has made a substantial contribution to the achievement of the Millennium Development Goals (MDGs) by reducing half of its poverty population.

Xiaoyi Wang
Editor

Ecological Migration and Targeted Poverty Alleviation in Ningxia

Experience and Lessons

SSAP

Springer

Editor
Xiaoyi Wang
Chinese Academy of Social Sciences
Beijing, China

Translated by
Sha She
Beijing, China

Xiaonan Zhang
Beijing, China

Sponsored by the Chinese Fund for the Humanities and Social Sciences

ISSN 2195-9781 ISSN 2195-979X (electronic)
International Research on Poverty Reduction
ISBN 978-981-19-7890-6 ISBN 978-981-19-7888-3 (eBook)
https://doi.org/10.1007/978-981-19-7888-3

Jointly published with Social Sciences Academic Press
The print edition is not for sale in China Mainland. Customers from China Mainland please order the print book from: Social Sciences Academic Press.

Translation from the Chinese language edition: "生态移民与精准扶贫: 宁夏的实践与经验" by Xiaoyi Wang, © Social Sciences Academic Press 2017. Published by Social Sciences Academic Press. All Rights Reserved.

This Springer imprint is published by the registered company Springer Nature Singapore Pte Ltd.
The registered company address is: 152 Beach Road, #21-01/04 Gateway East, Singapore 189721, Singapore

Foreword

Ecological Migration for Precision-Targeting Poverty Alleviation

Migration falls into types in China and other countries according to different criteria. Mass migration programs, whatever type, lead to social problems, and it is extremely difficult to succeed. Satisfaction and social integration of migrants, which are important measure of migration programs, are closely linked with their impact on migrant's living conditions and on the life of original residents of destinations.

China has set a goal to lift all rural residents living below the current poverty line out of poverty by 2020 when China will build a prosperous society in all respects. Its approaches to poverty eradication are summarized as "five measures" in poverty-stricken areas, namely boosting the economy, relocating the impoverished, carrying out ecological restoration, strengthening education, and improving social security. During the 12th Five-Year Plan period from 2011 to 2015, over eleven million impoverished people were relocated from ecologically fragile areas to areas of better conditions. During the 13th Five-Year Plan period from 2016 to 2020, China plans to relocate about ten million poor people. The ecological migration in Ningxia is a successful example in this regard.

Ningxia Hui autonomous region is an underdeveloped region in western China with a heavy concentration of poverty-stricken population, though its poor population is not the largest in the western region. At the end of 2015, 580,000 people in Ningxia lived below the poverty line, but the poverty headcount ratio in its rural area reached 14.5%, and 270,000 of these people were in Guyuan county. Xihaigu in Ningxia, and Dingxi and Hexi in Gansu used to be among the poorest areas in China. Since the 1980s, to alleviate poverty, Ningxia has been conducting relocation of whole villages to resettle impoverished people from ecologically fragile areas in the south to the Yellow River irrigation zone. By 2015, 1.16 million people in Ningxia, more than one-third of its rural population, had been relocated. Ningxia has completed an extremely challenging task, and has accumulated rich experience in the process. They have adopted multiple approaches to resettlement and have made constant efforts to improve relocation policies in order to provide migrants with ever-improving public utilities and services and more employment opportunities and raise their income. Ningxia's practice offers extensive experience in relocation for poverty alleviation.

The relocation programs for the purpose of poverty alleviation have been given different names, such as relocation and ecological migration. In essence, they aim to improve the lives of the impoverished by migration. The Institute of Sociology of Chinese Academy of Social Sciences (CASS), Ningxia Academy of Social Sciences, and North Minzu University conducted a joint survey of ecological migration in Ningxia in 2012, the results of which were published in *Ecological Migration, Development and Transformation: A Study of Migration and Poverty Reduction in Ningxia* by Springer. In the said book, we claimed that ecological migration is an effective measure to achieve dual goals of ecological protection and poverty alleviation. Ecological migration has provided the impoverished people in ecologically fragile areas with an opportunity to cast off poverty; therefore, it is well received by poor people in mountainous regions. Thanks to constant improvement of infrastructure and public services in resettlement areas, the migrants are very satisfied with the migration. Our survey also revealed problems resulting from the ecological migration, such as ecological deterioration in areas where people are resettled, social governance in settlements, and employment of certain groups of people after relocation.

The years between 2012 and 2016 are a critical period for implementation of the Outline for Development-Oriented Poverty Alleviation for China's Rural Areas (2011–2020). During those years, the ecological migration project in Ningxia relocated 3.45 million people, making it the largest scale of population relocation in the history of Ningxia. To check on the progress of ecological migration in Ningxia, analyze its social and economic impact, and draw experience of poverty alleviation by migration, the CASS organized a second survey of Ningxia's ecological migration. As with the first survey, this survey was conducted jointly by the Institute of Sociology of Chinese Academy of CASS, Ningxia Academy of Social Sciences, and North Minzu University. This research report is based on the results of the survey.

As revealed in the survey, governments at all levels in Ningxia have introduced new measures to facilitate the resettlement. They have adopted market mechanisms to promote participation of enterprises in economic development of settlements. They have also taken measures to increase the property income of the impoverished and turn poverty alleviation funds to property of poor people so as to ensure their stable income. They have made huge effort to create jobs for migrants, including creating public-service jobs for poverty-stricken migrants, developing labor intensive industries, and so forth. Thanks to the wide variety of poverty alleviation measures, the impoverished people have seen an increase in their income after relocation.

The survey also reveals that governments in Ningxia attach great importance to ecological protection and sustainable use of resources. Among others, they have been promoting water-saving irrigation techniques in relocation areas, and the bulk of land in places of migrants' origin has been transformed to ecological land. Ecological migration pursues two goals, i.e., poverty alleviation and ecological protection. Efforts are required to ensure ecological restoration in ecologically fragile origins and avoid ecological damages in destinations. Many destinations of ecological migration do not boast excellent ecology or abundant natural resources and have been chosen to receive ecological migrants after upgrading. Great effort is thus required to prevent ecological degradation due to the relocation campaign. Governments at all levels in Ningxia have taken actions to strengthen ecological protection in relocation areas.

Meanwhile, ecological migrants remain a fragile population, and the poverty headcount ratios in settlements are still high. People enjoy better conditions for agricultural production in the Yellow River irrigation zone, but still have limited agricultural resources and small plots of inferior farmland. The agriculture holds little potential for raising income. The migrants are also facing difficulties finding nonfarm jobs. Most of them can only find unstable, low-paying jobs, and old people find it hard to land nonfarm jobs. These issues must be addressed to prevent the people who have been lifted out of poverty from falling back into poverty.

It will be an extraordinary achievement in the history of mankind for China to eradicate absolute poverty in rural areas by 2020. Ecological migration is an important component of China's poverty alleviation campaign. We will keep following the progress of ecological migration in Ningxia to give an accurate account of the remarkable campaign and the changing lives of the impoverished.

Beijing, China
January 2017

Li Peilin

Contents

Chapter 1
Introduction Relocation and Targeted Poverty Alleviation: A Re-examination of Ecological Migration in Ningxia

Xiaoyi Wang

Relocation is an important measure in China to lift poverty-stricken people in inhospitable areas out of poverty. "Since 2012, the state has allocated 40.4 billion yuan from the central government to leverage a total investment of141.2 billion yuan of all kinds, resettling 5.91 million impoverished people. The central and provincial coffers as well as local governments at various levels have provided thirty-eight billion yuan for poverty reduction, resettled 5.8 million poor people. These concerted efforts have effectively brought more development opportunities to poverty-stricken areas".[1] China needs to lift over fifty-six million more people out of poverty during the 13th Five-year Plan period, and nearly ten million of these people need to be relocated. According to the 13th Five-year Plan for poverty alleviation by relocation issued by the National Development and Reform Commission (NDRC), China plans to relocate 9.81 million people in five years and introduce measures to lift the relocated people out of poverty.

Ningxia started relocation for the purpose of poverty alleviation in the 1980s. By the end of 2015, about 1.16 million people had been relocated from the ecologically fragile mountainous areas in the south of Ningxia. Most of these people were resettled in the northern and central parts of Ningxia, and a small portion was relocated to other mountainous areas in the south. According to the plan for poverty alleviation by relocation in the 13th Five-year Plan period, Ningxia needs to relocate only 80,000 people in the form of labor migration. Ningxia has completed the task of relocation for poverty alleviation. Its experience in this regard is thus useful for the rest of China to advance precision-targeting poverty alleviation.

[1] The State Council Information Office of the People's Republic of China. *China's Progress in Poverty Reduction and Human Rights*, 2016.

X. Wang (✉)
Chinese Academy of Social Sciences, Beijing, China
e-mail: xyang@cass.org.cn

X. Wang (ed.), *Ecological Migration and Targeted Poverty Alleviation in Ningxia*, International Research on Poverty Reduction,
https://doi.org/10.1007/978-981-19-7888-3_1

The relocation program has enabled the migrants to have access to better public services and higher income and has diversified the employment structure. The ecological environment in places of origin and destination has all shown significant improvement. But relocation does not necessarily lead to poverty eradication. Poverty headcount ratio remains high in relocation areas. Moreover, the relocation has also resulted in some new causes of poverty, which raises new challenges to the precision-targeting poverty alleviation campaign. In this section, based on our survey in the central and northern parts of Ningxia, we analyze the problems in anti-poverty governance in relocation areas and offer our advice on policy adjustment.

1.1 Policies on Relocation for Poverty Alleviation

China's poverty alleviation campaign started with its poverty alleviation efforts in the 1980s in Xihaigu in Ningxia and Dingxi and Hexi in Gansu. In Ningxia, the mountainous areas in the south and arid areas in the central part suffer from contiguous poverty, fragile ecosystem and severe shortage of resources for their population. In order to reduce the pressure on the environment and alleviate poverty, Ningxia has implemented five relocation projects (see Table 1.1), resettling 1.16 million impoverished people to the central and northern areas. Ningxia's poverty alleviation relocation projects last a long time and involve a substantial number of people. They have introduced a wide array of measures to resettle the migrants and improve their lives, thus creating the Ningxia model of relocation for poverty alleviation.

The relocation efforts in Ningxia started early in the 1980s and lasted over three decades. In the 13th Five-year Plan period, except for labor migration of a small number of people, Ningxia has completed its task of relocation. The government's spending on relocation has been increasing over the years. In the early stage of *Diaozhuang* migration, the relocated villagers were allowed to keep residences in both places of origin and destination and choose where to stay. With a close association between the origin and the destination, the relocation areas were actually expansion of the original villages.[2] In early *Diaozhuang* projects, government spending mainly went to public utilities, such as water conservancy projects and roads. The migrants needed to cultivate most of the land and build residences on their own and had great mobility.[3] Back then, per capita government spending on migration was less than 1000 yuan. In later relocation projects, government spending kept rising, and the infrastructure in relocation areas was also improving. During the ecological migration project in the 12th Five-year Plan period, the per capita government spending on migration reached nearly 30,000 yuan. The migrants were provided

[2] Shu (2015).

[3] In our survey, we found that early migrants had a vivid memory of the hardships in the early days of relocation. According to them, the relocation areas were so windy and dusty that there were grains of sand in their meals. Many migrants moved back to their hometowns. Thanks to land cultivation, afforestation and other projects to improve the conditions in villages, the ecology of the relocation areas is now much better.

Table 1.1 Five major relocation projects in Ningxia

Project	Period	Number of people relocated (unit: million)	Per capita spending (yuan)	Per capita income before relocation (yuan)	Per capita income in 2013 (yuan)
Diaozhuang migration (whole-village relocation)	1983–2000	1.98	931	60	7359
Yellow River lifting for irrigation	1998–2015	3.08	9059	948	5600
Relocation for poverty alleviation	2001–2007	1.53	9036	1,659	5443
Ecological migration in arid counties in the 11th Five-year plan period	2008–2012	15.36	18,000	2,235	6272
Ecological migration in the central and southern areas in the 12th Five-year plan period	2011–2015	3.46	28,800	N/A	N/A
Sum		115.86			

Note The data provided by Li Yaosong et al. of Ningxia Academy of Social Sciences are different. According to them, 3.45 million people were relocated in the *Diaozhuang* project, and 1.94 million people were relocated in Hongsipu alone in the Yellow River lifting for irrigation project. See: Li Yaosong et al., "On sustainable development of ecological migration in Ningxia," *Social Sciences in Ningxia*, 1(2012)

houses, and the settlements had essential facilities such as roads, schools and offices for villager committees. Water-saving irrigation facilities were adopted to improve the efficiency of water utilization.

Considering its total population, the migration implemented in Ningxia is a large-scale one. In 2015, Ningxia had a total population of 6.68 million, including three million rural residents. In that year, Ningxia relocated 1.16 million people, 38% of its rural population. In 2010, Ningxia's arid central area and mountainous southern area were priority areas for the state's poverty alleviation effort, with a population of 1.4 million, one million living in poverty. In the ecological migration project during the 12th Five-year Plan period, 35% of these impoverished people were relocated. The mass migration has had a significant impact on Ningxia's social and economic development. It has changed the distribution of population in Ningxia. According to the Statistics Bureau of Ningxia, at the end of the 12th Five-year Plan period, people living in the nine counties in mountainous areas in southern Ningxia accounted for 34.3% of its total population, and nearly two-thirds of its population lived in the plain in the north. In the 12th Five-year Plan period, the population in the plain in northern Ningxia increased by 1.48 percentage points.[4]

In implementation of the migration projects, the migrants in Ningxia are mainly resettled in newly built settlements, while a number of other measures are also adopted. The settlements are mainly situated in arid central Ningxia and in the north on the edge of the Maowusu Desert which used to be wasteland. Irrigation facilities and houses have been built in these places before the migrants are resettled. The settlements usually have a large population, some with one thousand households. Besides whole-village relocation, Ningxia also takes measures to resettle scattered migrants. Some migrants have chosen to resettle with their relatives or friends, and some are moved to existing villages with comparatively better natural conditions. Some relocated people, known as labor migrants, are resettled near townships in the north so that they can find nonfarm jobs. The labor migrants are not given land, but have access to social security as urban residents. In addition to people who have been relocated in projects initiated by the government, a large amount of people have also migrated on their own. These people are not resettled with government assistance. Instead, they have bought residences and land right of government-organized migrants.

The income of the migrants has increased rapidly. According to Ningxia Poverty Alleviation Department, the average income is over 5000 yuan for migrants in general and 7000 yuan for those relocated in the early *Diaozhuang* project. A great amount of money has been spent on land exploitation in migration projects, but the relocated people have made more income mainly because they have been migrant workers. According to a survey of 804 households in ten settlements in 2016,[5] before relocation, their income mainly came from plant production and animal husbandry, and only less than 10% of the respondents relied on migrant jobs as a major source of income; after relocation, income from migrant jobs has become a major source of income for over 60% of the respondents. Now that the migrants are closer to cities in northern Ningxia and other provinces and have access to better transport facilities, and they can find more nonfarm jobs. Short-term jobs and nonfarm jobs have also increased in places near relocation areas. Women and middle-aged and old rural residents in relocation areas have found nonfarm jobs, such as picking goji berries and planting grapes, mainly in agricultural enterprises near their villages.

As public utilities and services in relocation areas continue to improve and migrant's income keeps rising, the migration projects hold increasing appeal to the impoverished people in mountainous areas in southern Ningxia. Nearly 50% of the 804 respondents moved to settlements on their own, instead of being required by the government. Some migrants who have lived in relocation areas for a long time told us that the conditions in settlements are now much better than in their hometowns. Settlements have good transport facilities, while their hometowns are deep in the

[4] Statistics Bureau of Ningxia Hui Autonomous Region, *Report on Economic and Social Development of Ningxia during the 12th Five-year Plan period (15)*, April 2016.

[5] The survey was conducted by the Institute of Sociology of CASS and North Minzu University in 2016.

mountains, with poor transport facilities. Migrants also have access to better farmland and irrigation facilities in relocation areas. The ever-improving public services are the most important factor that appeals to migrants. Almost all migrants believe that the settlements are more convenient than their hometowns. Better education in relocation areas is also a major factor that appeals to migrants. Meanwhile, migrants are involved in urbanization. Emigration of farmers is an inevitable trend whether or not the ecological migration projects have been implemented. Against such a backdrop, settlements have become a springboard bolstering the emigration from mountains in southern Ningxia. Some relocated villagers have been migrant workers in places near settlements and have left the villages for urban areas after getting accustomed to urban life. In settlements where people are offered jobs instead of land, despite unstable employment status, most migrants do not want to forfeit their land for urban household registration. Though, all respondents said that they would find jobs in townships and that they had never thought of returning to the mountainous areas in southern Ningxia even considering the decreasing job opportunities in urban areas in recent years. Moreover, it is hard for farmers to return to their original villages, because there has been no more efforts to build public utilities in those villages since the relocation, and houses in some villages have been demolished. Due to lack of public services, these villages no longer appeal to people.

Due to unbalanced distribution of job opportunities and public services, outmigration has become an inevitable trend. Whether or not the migration projects have been implemented, rural residents will inevitably migrate from mountains to plains and urban areas. The migration projects have been a success because they follow the trend. These projects provide farmers with public utilities and services, thus reducing their migration cost. Meanwhile, as migrants build new homes in places of better geographical conditions, they can still work in agriculture, which reduces the risks of migration. On the other hand, the relocation means that migrants must leave their hometowns, which can mean different things for different groups of people. For labor-abundant households, migration presents more opportunities. Impoverished households with inadequate labor, however, can make only limited use of the opportunities arising from migration.

1.2 Relocation and Poverty

Relocation does not necessarily lead to poverty elimination. Poverty remains a major issue in many settlements. Settlements in the ecological migration project in the 12th Five-year Plan period have high poverty headcount ratios. Among the ten settlements covered in the survey conducted by the Institute of Sociology of CASS and North Minzu University, eight villages are registered as impoverished. These eight villages have 10,436 households, and 18% of the total, or 1860 households live in poverty. In four of these villages, poverty-stricken households account for over 35% of the total (see Table 1.2). The poverty headcount ratio in settlements built in the ecological migration project in 12th Five-year Plan period is even higher. Three of the ten

Table 1.2 Poverty in villages covered in the survey

Village	Year of relocation	Number of households	Poverty-stricken households	Percentage of poverty-stricken households
Qihai	2012	887	429	48
Xinghai	2012	612	312	51
Yuanlong	2012	1983	73	4
Hongya	2003	554	206	37
Lihua	2000	721	55	8
Chaoyang	1999	1420	653	46
Dahe	1998	1296	110	8
Tuanjie	1996	658	10	2
Xinhua	1996	1502	12	1
Mulan	1994	803	0	0

Unit Household, %
Note The data are based on village-level survey. The percentage of poverty-stricken households as revealed in the household-based survey is 14%, lower than the figure based on village-level survey. The discrepancy is probably caused by sampling bias
Source of data Survey conducted by the Institute of Sociology of CASS and North Minzu University in 2016

villages in the survey were established in the ecological migration project in the 12th Five-year Plan period, and poverty-stricken households in two villages account for nearly 50% of the total. Yueyahu, a town in Yinchuan city of Ningxia, is a typical relocation community. In implementation of the precision-targeting poverty alleviation strategy, all five newly established settlements have been determined as poverty-stricken ones, while all seven old settlements have not. Moreover, in most of the five new settlements, over 50% of the households live in poverty. In theory, migrants are relocated from ecologically fragile areas to places with better conditions, thus having access to better resources. In practice, however, the issue is much more complicated. In many settlements, the migrants have not seen increase in available agricultural resources as a result of the migration project. What's worse is that the resources available may have reduced. For example, they may have less arable land and do not have any grassland or wasteland to cultivate. Some irrigated farmland has not been effectively utilized. Owing to these problems, after the mass relocation for poverty alleviation in Ningxia, poverty has been a common problem for residents of Hongsipu, a community of migrants. Hongsipu has been made a priority area in poverty alleviation in Ningxia and the ninth county on Ningxia's list of poverty-stricken counties.[6]

The high poverty rate in settlements established in the ecological migration project in the 12th Five-year Plan period is attributable to two reasons. First, new migrants

[6] Wu Zhandong, *Report on Implementation of the Ecological Migration Plan in Central and Southern Ningxia in the 12th Five-year Plan Period*, December 2011.

have not yet accustomed themselves to life in the new environment. Though many settlements were established in 2012, most of the villagers were resettled in these villages after 2012 and some even after 2016. It takes time for them to adapt to the new environment, including learning about local job market, creating their social network and learning new skills. Before that, they are not competitive in the labor market. Second, as new relocation policies are introduced, it is more difficult to lift the migrants out of poverty. Considering the ever-decreasing water resources and arable land, each of the ecological migrants in the 12th Five-year Plan period is granted only 1 *mu* (about 0.07 ha) of farmland. The farmland is not operated by individual households, but leased by village-level cooperatives to enterprises for operation. The migrant households receive 300–500 yuan a year from those enterprises for leasing their farmland, which is far from inadequate.[7] Early migrants usually have more farmland. According to the then policies, they received more than 2 *mu* (about 0.13 ha) of farmland cultivated by the government. Most of them have also cultivated more land on their own. In Haitaobei, one of the earliest settlements in Ningxia, for example, the migrants have been cultivating the wasteland since they were resettled here, and now each household in the village has about 2 ha of farmland and a spacious courtyard. One of the respondents in our survey has a courtyard of about 2667 m^2, in which grapes and other fruit trees are planted for courtyard economy. Villagers in Dahe have also expanded their farmland through constant cultivation of nearby wasteland. Old migrants receive income from multiple sources including agriculture, animal husbandry and working in urban areas and have thus made more income. New migrants, however, do not have any farmland and can only live on income from working in urban areas and government allowances, thus receiving lower income. In addition, new migrants have shown limited capability to fend off risks.

Besides high poverty rate in new relocation areas, some old relocation areas have also fallen into poverty due to desertification and salinization. Hongya and Chaoyang in Table 1.2, for example, are both old settlements with a poverty rate over 1/3. Migration projects aim to improve the living conditions of the impoverished people by relocating them from ecologically fragile areas. But the relocation destinations usually do not have ideal conditions and require substantial input of labor and resources for transformation. The ecosystem in places of relocation destination in Ningxia is also fragile. While the migration has mitigated the pressure on the environment in southern Ningxia, it has increased such pressure in places of destination and resulted in shortage of water and desertification in these places.[8] Due to low rainfall and rapid evaporation, many relocation areas, especially the arid area in the central part of Ningxia, suffer from shortage of water.[9] Hongsipu, with a heavy concentration of migrants, is confronted with lack of irrigation water, desertification and salinization.

[7] The farmland is leased to enterprises for farming for two reasons. First, the newly cultivated land is hard to farm for individual households because of poor land quality and inadequate facilities. Second, extensive land and operation by enterprises are essential for industrialization of agriculture. As a matter of fact, the migrants cannot make ends meet with the small area of farmland granted to them.

[8] Du et al. (2012).

[9] Li (2015).

Water shortage has had a strong impact on agricultural production in Hongya village. Farmers used to irrigate their corn four to five times during the growing season, but only once or twice now. Corn yield has reduced by 2/5 due to lack of water in general, and some plots have no harvest at all. Local villagers believe that water shortage is mainly caused by increasing cultivation of land by farmers and agricultural enterprises. Among the thirty-eight relocation areas for migrants in Ningxia, twenty-five areas are confronted with shortage of water.[10] While it is not feasible to increase the total amount of irrigation water, water-saving irrigation techniques are essential to expanding the area of arable land. Such techniques can reduce the water used for irrigation. But it remains unclear what impact the decrease in irrigation water has on the ecology of relocation areas.

While some areas do not have sufficient water for irrigation, some other areas face problems due to poor drainage. Desertification and salinization thus concur in Ningxia, especially in the arid areas in central Ningxia. Measures have been adopted to contain desertification and salinization, but with no significant effect. Moreover, due to high cost involved in implementation of such measures, it is hard to win the support of migrants. In Dahe village, for example, over 30% of its farmland has become saline and thus not suitable for growing crops. Soil salinity can be reduced by building drainage and ditching the plot, but according to local villagers, the effect hasn't been satisfactory. Goji berries can now grow in farmland in which salinity is not very high. But if salinization cannot be contained effectively, people will have to abandon the farmland of extremely high salinity.

With an extensive area of wasteland to be cultivated and the Yellow River for irrigation, Ningxia has developed a great amount of farmland. But similar to migrants in other parts of China, most of the migrants in Ningxia do not have adequate agricultural resources. With small plots of arable land of poor quality, the relocated households can barely make ends meet. In addition, they face great difficulties changing the planting structure. Villagers of Hongya told us that they had made eight adjustments to the village's industrial structure, all to no avail. Farming of cash crops such as goji berries and grapes is thriving. But a survey in Hongsipu reveals that most migrants are still growing corn, despite fewer subsidies from the government. Farmers prefer low-profit staple crops because of lower cost, simple techniques required, market and so on. In theory, farmer cooperatives can help farmers deal with problems relating to cost, techniques and market in production of cash crops. According to our survey, however, most cooperatives have failed to perform such functions. In new relocation areas, cooperatives have only leased land to enterprises on behalf of villagers.

For most migrants, it is fairly difficult to cast off poverty by agricultural production. According to the 804 questionnaires, we received in the survey agricultural income accounts for no more than 25% of the household income of the respondents, while most of their income is from working for enterprises. The average household income of the respondents exceeds 320,000 yuan, with only eight thousand yuan from agricultural production. Of the 804 respondents, 409 people have agricultural income, and nearly 60% of these households earn no more than 10,000 yuan from

---

[10] Li (2015).

agricultural production. In another word, 20% of these migrants have agricultural income of over 10,000 yuan.

According to the 13th Five-year Plan for relocation, migration projects are mainly implemented in inhospitable areas such as those deep in the mountains, areas where development is limited or prohibited according to national plans, areas with inadequate infrastructure where it is too costly to build and operate public facilities and areas with high rates of endemic diseases and geohazards. Migration can significantly improve the production conditions of migrants. In practice, however, the places of destination are also ecologically fragile, without adequate resources for agricultural production. As most of the relocated people do not have access to adequate agricultural resources to make ends meet, they prefer to work as migrant workers to make more income. For 50% of the 426 respondents who have been migrant workers, their income from working out of hometown accounts for more than 75% of the household income. Though the migration projects have enabled people to find nonfarm jobs in closer places, according to our survey, the migrants cannot make stable income as migrant workers. As the migrants lack necessary skills and training, they mostly do low-paying jobs such as odd jobs at construction sites, working in road construction projects or in coal miners in Inner Mongolia. Some migrants earn a living by picking goji berries and grapes in neighboring areas or Qinghai.

Most migrants cannot find stable nonfarm jobs. Take the seasonal job of picking goji berries for example. Workers are only needed in about three months a year in Qinghai and a bit longer in neighboring areas to pick goji berries. Low-skilled jobs at construction sites are not stable either. As no specific skills are required to do such jobs, workers tend to be replaced. Contractors bring the migrants together when workers are needed, but once the work is finished, they have to find new jobs. Generally, Miaomiaohu villagers can only find jobs in about five months in a year. In addition, the overall economy has a significant impact on the market of low-skilled jobs. Over the recent years, with decreasing investment in the real estate industry and infrastructure, migrants have been confronted with ever-increasing difficulties in finding jobs, and their wages have been declining. For some people doing odd jobs at construction sites, their daily wage is 1/3 lower, and their work time has reduced by 1/3 or 1/2.[11] Low-skilled nonfarm jobs are not open to middle-aged and elderly migrants and women, making their livelihoods vulnerable. Our survey of the migrants shows that male migrants aged over forty-five face considerable difficulties finding jobs. In traditional rural communities, people aged over fifty can return to villages to work in agriculture if they are unable to find jobs in urban areas. For the bulk of the migrants, however, chances are small that they can make a living on agricultural production. Early migrants have more farmland, but they receive small income from agricultural activities due to water shortage and poor land quality. Later migrants have only small areas of farmland. A common problem for all migrants is that they only have farmland, but do not have any other agricultural resources, such

[11] According to some migrants covered in the survey, their daily wages, which used to be 120–150 yuan, had dropped to 80–100 yuan. Previously, they could find odd jobs in 20-plus days every month. Now they could only find jobs in 15 days a month at the best.

as grassland and hills for pasturage and mountain slopes where they can grow trees. Unable to live on agriculture and find nonfarm jobs, many migrants over fifty have been living in poverty. This is also the case for women. While new labor-intensive enterprises have created some additional jobs for women, most women can only find seasonal jobs. Unstable employment has resulted in fluctuation in migrant's income. As a result, families without adequate young and middle-aged labor force are vulnerable to poverty.

Relocation has created opportunities for many migrants, but it has not addressed the root causes of poverty in rural areas. Families without adequate labor force are still vulnerable to poverty. Unlike previously, as the migrants have fewer opportunities to make income from agricultural activities, they find it difficult to make ends meet by agricultural production when they cannot find nonfarm jobs. In a sense, relocation has increased the vulnerability of migrants.

Relocation projects are both self-initiated and compulsory. By addressing these projects as relocation and ecological migration, we are actually emphasizing different facets of the migration. On the one hand, we need to relocate people from ecologically fragile areas so as to lift them out of poverty. On the other hand, people must be moved out of ecologically fragile areas in order to restore ecology in these areas. Despite different priorities, these two facets have been addressed simultaneously in practice. Ningxia's ecological migration project in the 12th Five-year Plan period is a good example in this regard. People are relocated from ecologically fragile areas for ecological protection, and efforts are made to improve the living and production conditions of the migrants. The dual goals, however, have caused problems in the migration. Relocation is compulsory, because all residents in ecologically fragile areas, poor and rich, whether they are ready to migrate, are required to be relocated for the purpose of ecological restoration. The migrants are given certain options in means to resettlement, but they have not been compensated based on ecological service value. Instead, they are supported in development by means of development-oriented poverty alleviation. Most resettlement arrangements of the migrants are not conducted on a self-initiated basis, and the migrants are allowed limited options. Ecological migrants have not received adequate compensation for their property. Migrants of better economic and social conditions with more human capital have benefited from the migration. Poverty-stricken households without adequate capital, however, still live in poverty, who are the objects of precision-targeting poverty alleviation.

The migrants have experienced limited benefit from the migration in terms of agriculture, non-agriculture industries and compensation, which is the main cause of high poverty rate in relocation areas, especially in new relocation areas. Poverty-stricken households usually do not have adequate people who are able to work.

1.3 Targeted Poverty Alleviation in Relocation Areas

With an accurate understanding of the poverty issue in migration, Ningxia Hui autonomous region has adopted a lot of measures to effectively alleviate poverty in a targeted manner. Ningxia government has attached great importance to poverty alleviation in settlements, as reflected in the high percentage of poverty-stricken households in settlements established in the 12th Five-year plan period. They have introduced policies and measures to increase government spending on relocation projects, boost nonfarm employment, raise property income and improve the ecological environment in relocation areas. Due to weak self-organizing capacity and frequent residential mobility of settlements, they have been confronted with challenges in precision-targeting poverty alleviation in relocation areas.

To help the relocated residents settle down and adapt to the new environment, Ningxia has introduced policies mainly in three areas. First, before expiration of the relevant preferential policies introduced before the relocation, the migrants shall continue to enjoy such preferential policies, such as subsidies for returning farmland to forest, direct subsidies to grain producers and so on. These policies have not ceased because of the relocation or withdrawal of farmland. Second, improve the infrastructure in relocation areas, including public facilities, roads and rural housing. Third, cultivate the land for farming, build water conservancy facilities, and fund farmers in production of cash crops. These policies have facilitated the migrant's adaptation to their new life and agricultural production. Such general policies have facilitated the development of most migrants. Nonetheless, they have had limited effect on the poverty-stricken migrants.

Nonfarm employment is a major means for migrants to achieve development.

Our survey in 2012 showed that due to inadequate labor-intensive industries in the western region of China, nonfarm industries could provide only limited jobs. Every 10,000 yuan worth of industrial output create fewer jobs in Ningxia than Guangdong, Zhejiang and other developed regions in the east of China.[12] Some labor-intensive industries are coming to Ningxia. The newly established spinning mill in Dawukou, for example, has played a significant role in boosting employment of migrants. Now, unlike middle-aged and elderly migrants who can only find seasonal jobs at construction sites, young migrants have more opportunities to find stable jobs in enterprises. In relocation communities in Dawukou where the residents are offered jobs instead of land, some migrants, after training, have been able to find jobs in transportation and other industries that require professional skills. Poverty-stricken households are given preference in application for public-service jobs, which has increased their opportunities of finding a job. In Binhe New district, for example, poor migrants are given preference in application for cleaning jobs. Meanwhile, as Ningxia makes progress in agricultural industrialization, agricultural enterprises have also created a lot of jobs for the migrants. While villages established in the 12th Five-year Plan migration project transferred their farmland to agricultural enterprises,

[12] Li and Wang (2013).

these companies usually promised to give preference to the villagers when they hire employees.

In addition to measures directly supporting the industrial development in relocation areas, efforts are also made to boost farmer's property income by means of collaborative operation. Transfer of land rights from the ecological migrants relocated in the 12th Five-year Plan period is a way to generate property income. Farmers do not work on their farmland, but transfer the land rights to agricultural enterprises at a certain price. Farmers also earn profits from beef and dairy cattle purchased with government's poverty alleviation funds which are tended by enterprises, which helps increase their property income. Cattle management by enterprises has enabled standard and procedure-based cattle raising and the production of quality beef and milk. Meanwhile, these enterprises provide the migrants with fixed dividend, which has raised the migrant's income and reduced the operation risks of farmers. Some settlements have initiated photovoltaic projects in which photovoltaic enterprises place photovoltaic panels in courtyards and on roofs of villagers, and farmers purchase shares of photovoltaic companies with poverty alleviation funds so that they can receive dividends on an annual basis.[13] Ningxia's experience of alleviating poverty with government funds shows that sound policies are required to transform such funds into property of poverty-stricken households and ensure their sustainable income. Beef cattle and dairy cattle companies, for example, have enjoyed preferential policies of local governments on loans, land use and others in management of poverty-stricken household's cattle. Photovoltaic projects for poverty alleviation are also given preference in approval, land use and others. Poverty alleviation based on property means more than allowing poverty-stricken households to purchase shares of enterprises. It requires concerted efforts of the government.

The government has also directed more funds to improve the quality of farmland in relocation areas and prevent desertification and salinization. Relevant authorities have been addressing soil salinization in relocation areas, and government spending in this regard has been on the rise. In 2014, the Development and Reform Commission of Ningxia Hui Autonomous Region spent 10.9 billion yuan addressing desertification and salinization, and the Department of Water Conservancy spent forty-five million yuan addressing desertification and salinization in about 2000 ha of farmland. These efforts have improved the conditions for agricultural production in some relocation areas. In Dahe village, for example, the soil salinity of some farmland is reduced by building drainage and ditching the plot.[14] These measures have mitigated the

[13] According to Su Zhijun, a respondent in Minning town, the cattle in Yuanlong village are managed by a company. Villagers buy shares of the company at the price of 2000 yuan per head of cattle and receive an annual dividend of 2000 yuan. Poverty-stricken households are funded by government's poverty alleviation funds in purchasing the company's shares. The villagers have also bought shares of a photovoltaic company with a subsidized loan of 100,000 yuan and receive an annual dividend of 15,000 yuan from the company. 5000 yuan of the dividend are used to repay the loan, and the remaining 10,000 yuan are paid to villagers.

[14] In our survey, we found that most of the villagers refused to give up their farmland for ditching, although they agreed that digging ditches was the best way to deal with salinization. Drain pipes are thus buried in the farmland to reduce salinity, but the effect has not been so good.

difficulties of some migrants, but fail to address the root cause of poverty of the migrants. While this is partly due to the fact that the government spending in this regard has not covered all relocation areas, it is also attributable to some special social governance issues in precision-targeting poverty alleviation in relocation areas.

More financial input is required in precision-targeting poverty alleviation. More importantly, not a single poverty-stricken household or individual shall be left behind in the campaign. To this end, we must identify all the poor and give them help and support. Due to great mobility of the migrants in relocation areas, it has been difficult to identify the poor. Despite government policies to promote population stabilization,[15] the migrants remain highly mobile. North Minzu University has studied the low resettlement rate of ecological migrants in Ningxia,[16] which reveals four causes of such low rate. First, some households which are required to be relocated have not moved as planned. Second, some households have resettled in settlements but have left later. Third, some households have resettled in settlements but have been working in other places. Fourth, some people are not required by the government to relocate but have chosen to migrate by purchasing or renting housing of government-planned migrants.

Population mobility is common in settlements. In Dahe village, for example, 50% of its over eight hundred households did not migrate to this place when the village was newly established. A large number of the villagers moved to Dahe of their own free will after a strict household registration system was put in place in 2005. As they do not have *hukou*, or household registration, in Dahe, they do not have access to preferential policies of poverty alleviation. According to head of the fourth villager group of Dahe, among the 120 households of the group, fifty households do not have household registration, fifty households with household registration live in the village, and twenty households with household registration have been moving around. The group head said that a minority of households had made more income and moved to their urban residence, but the majority were probably going from place to place. The situation in Hongya village is similar. According to officials of the village, the population of the village has been changing. Only 130-plus households of the initially resettled 470-plus ones have remained in the village, while others have moved back to their hometown or migrated to other places. Most of the households in Hongya are not resettled in the village at the very beginning, but have moved here in succession. As transactions of houses and farmland are prevalent in settlements, there has been a pricing mechanism for houses and farmland. In Hongsipu, a migrant's residence and 0.67 ha of farmland cost about 150,000 yuan.

A mechanism of population migration has taken shape in old settlements. In new settlements, however, some households which should have migrated have not done so. For example, 2800 households were intended to be resettled in Maqu of Hongsipu,

[15] Government policies prohibit the migrants from unlicensed transfer of farmland and housing, and severe punishments will be inflicted in case of violation. Local governments must withdraw any housing for ecological migrants and labor migrants and impose a fine if the housing is not occupied by these migrants or rented to others.

[16] North Minzu University, "A study of the low resettlement rate in Ningxia's relocation communities of ecological migrants," (2016).

but only 800-plus households are now living in this place. Permanent residents of Maqu account for merely 28% of the people who are supposed to be resettled. Half of the other nearly two thousand households have not moved out of their hometown, and half have moved away after being relocated in Maqu. Missing migrants are found in all relocation areas, but the percentages vary because of their different conditions. The Yueya Lake relocation area, for example, benefits from proximity to Yinchuan and has seen fewer missing migrants.

Swinging has helped the migrants adapt to the new environment, as is the case in the *Diaozhuang* project in Ningxia. Allowing migrants to swing between their hometowns and relocation areas has provided them both time and space to make a choice of relocation time and destination as they deem proper. With the increase of nonfarm jobs and the progress in rural urbanization, the migrants are not only swinging between their hometowns and relocation areas but are also likely to move to urban areas and other places.[17] As the targeted measures to alleviate poverty are based on specific residents, the swinging population have caused difficulties for precision-targeting poverty alleviation. Since a strict household registration system for migrants is put in place in 2005, those who are not required to migrate but have resettled in relocation areas out of their free will have found it difficult to obtain household registration in these areas, causing inconsistency between registered permanent residence and actual residence of residents in settlements. Residents who have moved to relocation areas out of their free will after 2005 do not have registered permanent residence in these areas, while some planned migrants, with registered permanent residence in settlements, do not live in these villages. Some people who are required to be relocated have not moved to settlements, and their registered permanent residences are not in these villages. Only a proportion of the residents in settlements are permanent residents with registered permanent residence in these villages. According to the policies of precision-targeting poverty alleviation, people without registered permanent residence in settlements are usually not covered by precision-targeting poverty alleviation measures in these villages. Officials of two villages, Dahe and Hongya, both mentioned that poverty alleviation policies aimed at villagers with household registration in the villages. According to them, residents without registered permanent residence in the villages did not have access to subsistence allowances and could not be registered as poverty-stricken households. As for people with registered permanent residence in settlements but not living in these villages, they are, theoretically, the objects of poverty alleviation policies. In practice, however, it is difficult to include them in poverty alleviation efforts because they have been away from the villages and their situation remains unclear to relevant authorities. Officials of Dahe village told us that they helped the impoverished people without registered permanent residence in the village by other means, but subsistence allowance and registration of poverty-stricken households must be based on registered permanent residence. Targeted poverty alleviation policies, which are based on villages, fail to consider the great mobility of migrants. Consequently, some impoverished people have not been covered by such policies.

[17] Wang (2011).

It has been a challenging task to precisely identify the poor in settlements, especially in new settlements. This is due to several reasons. First, some measures found effective in other places have not been as helpful in settlements. For example, since the housing conditions of impoverished farmers are usually poor, housing condition has been an important criterion for identifying the poor. In new settlements, however, the houses are built by the government according to unified standards, thus housing conditions cannot help identify the poor. Second, as residents of settlements are not familiar with each other, they are under lighter moral pressure to be honest in applying for status of poverty-stricken households. Most villagers want to be identified as poverty-stricken households. Such unfamiliarity also makes it more difficult to obtain accurate information than in other villages. Participation of villagers has been a major measure to better identify the poor, but it has not been as effective in settlements where villagers are not familiar with each other. Officials of settlements must conduct door-to-door interviews to ensure the results of poverty identification.[18]

Due to high poverty rate, great population mobility and difficulties in identification of the poor, three problems tend to arise in the process of relocation. First, while attaching great importance to giving full play to the role of local governments, inadequate attention has been paid to arousing farmer's initiative. As a result of inadequate participation of migrants and demanding tasks of poverty alleviation, local governments have played a crucial role in poverty alleviation in relocation areas. The Party secretary of Yuanlong village is sent by the Minning town government. Many agricultural enterprises are established in relocation areas in Maqu thanks to the efforts of local governments. Second, governments tends to take all matters into their hands and replace development-oriented poverty relief measures with disguised subsidies. This is especially the case in some property-based poverty alleviation projects which are organized and implemented by the government, while the impoverished households merely receive dividends without actively participating in such projects. Property-based poverty alleviation measures have thus turned into disguised subsidies. Third, the coverage of poverty alleviation measures has been expanding, but some impoverished households do not have access to such measures. Thanks to wide coverage of poverty alleviation measures, it is easier to include the migrants with registered permanent residence in settlements and living in these villages into poverty relief projects. But people who do not have registered permanent residence in settlements or who have been away from the villages for a long time are not covered by poverty alleviation measures, resulting in omissions in the precision-targeting poverty alleviation campaign.

[18] A respondent in Mingning said: "Yuanlong is a newly established settlement. We are not familiar with the villagers, and have to conduct door-to-door interviews time and again. An old man who has several grandchildren was initially registered as impoverished. While visiting him, we paid attention to his residence and clothing, and asked if he had any income. He said no, but we later checked his social security record and found he was a retired teacher receiving over 5000 yuan in pension each month. We must go to specific households concerned to verify, sometimes more than once, and ask neighbors for information. A problem is that the villagers are not familiar with each other. We have to check time and again.".

1.4 Targeted Poverty Alleviation and Adaption to Mobility

The relocation projects have two goals to meet, i.e., poverty alleviation and ecological protection. The government intends to resettle the migrants in designated areas, but migrants have shown great mobility. Measures are taken to boost nonfarm employment, but some migrants have been unable to land jobs. All these have caused some major problems in relocation for poverty alleviation. In alleviating poverty by relocation, we must pay attention to expanding the property of farmers and establishing a mechanism for precision-targeting poverty alleviation that adapts to the great mobility of migrants. Meanwhile, we must strengthen the social support to excluded population and establish a poverty alleviation mechanism based on mobility to replace the existing relocation scheme mainly in the form of establishing new settlements.

A poverty alleviation mechanism based on mobility will break the dual structure of moving in and out. While not changing the ownership of natural resources in ecologically fragile areas, ecological compensation is introduced to boost farmer's income and protect the environment. In addition to measures severing the connection between migrants and their original villages, efforts are also required to encourage migrant's participation in environmental protection and promote various types of employment. The number of settlements should be cut significantly, and rural–urban integration must be promoted to support migrants in finding nonfarm jobs and boost urbanization.

First, different policies should be adopted for ecological migration and poverty-alleviation-oriented migration. Migration for poverty alleviation concurrents with ecological migration, but the migration policies must be improved in view of their respective characteristics. Policies on ecological migration should focus on ecological compensation for all migrants based on ecological service value while acknowledging their contribution to ecological protection. In Ningxia, the ecological migrants in the 12th Five-year Plan period have given their housing and farmland to the government for ecological restoration, but such ecological assets have not been evaluated, and the migrants have not received compensation for their ecological assets. The government must pay ecological compensation to migrants based on relevant policies on ecological compensation such as subsidies for returning farmland to forest, non-commercial forest, awards and subsidies for grassland protection. Policies on migration for poverty alleviation are aimed at helping poverty-stricken households of the ecological migrants, or, in another word, rural households living in poverty after receiving ecological compensation. With a clear understanding of the compensation for ecological migrants and relocation for poverty alleviation, we must prevent migrants from falling into poverty due to inadequate compensation and avoid inclusion of ineligible people in poverty alleviation projects.

Second, we must build a poverty alleviation mechanism that adapts to the great mobility of migrants. As increasing mobility of rural population has become an inevitable trend in China, resettlement of migrants in designated relocation areas can no longer satisfy the needs of migrants and may lead to further environmental and social issues. Migrants are looking for a proper way of life while moving around. Therefore, poverty alleviation policies, instead of trying to reduce the mobility of migrants, must adapt to such mobility. While advancing rural–urban integration, constant efforts are required to improve the poverty alleviation mechanism so that poverty-stricken migrants can also benefit from public services in urban areas.

Third, in view of the exclusion of middle-aged and old people in the labor market, efforts are required to promote informal employment and improve the social security system. Some migrants and middle-aged and old farmers having difficulties landing a job should be allowed and encouraged to participate in ecological protection projects in the mountainous region in southern Ningxia, which will increase their income. Measures should be taken to improve agricultural production conditions in relocation areas and support farmers in development of family breeding farms so that middle-aged and elderly migrants can engage in agricultural activities. Policies should be introduced to encourage enterprises to employ middle-aged rural residents.

After years of relocation efforts, the number of people living in inhospitable areas has dropped significantly. The migration projects are aimed at not only relocating people from inhospitable areas, but also reducing the pressure of population in ecologically fragile areas and improving the living conditions in these areas. Therefore, we must gradually stop the one-size-fits-all approach in relocation and promote population mobility that can boost the development of ecologically fragile areas.

References

Du, H., et al. (2012). Study on the sustainable development of ecological migrants in Ningxia. *Journal of Anhui Agricultural Sciences, 40.*

Li, D. (2015). Problems in resettlement of ecological migrants in central and southern Ningxia and Solutions. *Journal of the Yinchuan Municipal Party College of C.P.C, 1.*

Li, P., & Wang, X. (2013). Migration, poverty alleviation and ecological progress: survey report on ecological migration in Ningxia. *Social Sciences in Ningxia, 3.*

Shu, X. (2015). Review and prospect of ecological immigration development course in Ningxia. *Journal of Ningxia Communist Party Institute, 2.*

Wang, X. (2011). From swinging to mobility: adaption in the migration. *Journal of Jiangsu Administration Institute, 6.*

Chapter 2
Brief Introduction of Socioeconomic Conditions of Ecological Migrants in Ningxia

Ye Fan and Hao Chen

The government of Ningxia Hui autonomous region started organizing mass migration in the early 1980s to ensure adequate food and clothing for people in contiguous poverty-stricken areas in Ningxia. Since 1983, Ningxia has implemented three major migration projects, i.e., the *Diaozhuang* project, the "1236" project (the migration for Yellow River Irrigation Project), and the ecological migration project. All these projects aim at development-oriented poverty alleviation and ecological restoration and fall into the scope of ecological migration in a broad sense. Ecological migration has enabled most migrants to have adequate food and clothing and earn higher income. Meanwhile, as the land withdrawn from migrants has all been used for ecological conservation, ecological migration has significantly reduced the population pressure on the environment and contained ecological deterioration in the areas of origin. By ecological migration, Ningxia has both boosted economic development and improved the environment.

2.1 Living Conditions of Ecological Migrants

Based on a survey of ten settlements in 2016, this section describes and analyzes the living conditions of migrants, including natural conditions, transport facilities, water supply, energy for household use, and public facilities.

Y. Fan (✉)
School of Economics, North Minzu University, Yinchuan, China
e-mail: 303844609@qq.com

H. Chen
North Minzu University, Yinchuan, China

X. Wang (ed.), *Ecological Migration and Targeted Poverty Alleviation in Ningxia*, International Research on Poverty Reduction,
https://doi.org/10.1007/978-981-19-7888-3_2

Table 2.1 Natural disasters in settlements (%)

Disaster	Share
Drought	40
Plant pest and disease	20
Animal epidemic	10
Earthquake	30
Sand and dust storm	40
Flood	0
Freezing rain/snowstorm	0
Dry hot wind	0
Mudslide and slide	0
Effective sample	10

2.1.1 Natural Conditions

Among the ten villages established as a result of ecological migration, 90% are in the rural area, and 10% are in the suburbs. According to data about the natural disasters they have suffered during the past five years, due to low rainfall, settlements are often affected by drought and sand and dust storms. Of all the settlements in Ningxia, 40% have been affected by drought, 20% have suffered from plant pests and diseases, 10% have had animal epidemics, 30% have experienced earthquakes, and 40% have had sand and dust storms (see Table 2.1).

2.1.2 Transport Facilities, Water Supply, and Energy for Household Use

After relocation from the mountainous region, the migrants live a more convenient life settlements have convenient access to transport. Of the roads in settlements, 70% are concrete roads and 30% are asphalt roads.

The quality of water for household use is also significantly improved. Many villagers could only rely on rainwater for household use, and some had to purchase water. Now, all settlements have clean and healthy tap water. Tap water has become the main source of drinking water in the ten villages.

A significant change has taken place in terms of household cooking fuel. The migrants, before relocation, mainly used firewood. Now, 50% of the settlements use electricity as the main cooking fuel, 20% use coal and electricity, and 10% use mainly coal. Firewood remains the main cooking fuel in only 20% of the villages (see Fig. 2.1). As clean and efficient new energy is gaining popularity, 50% of the settlements have started using solar energy, and 40% have started using biogas.

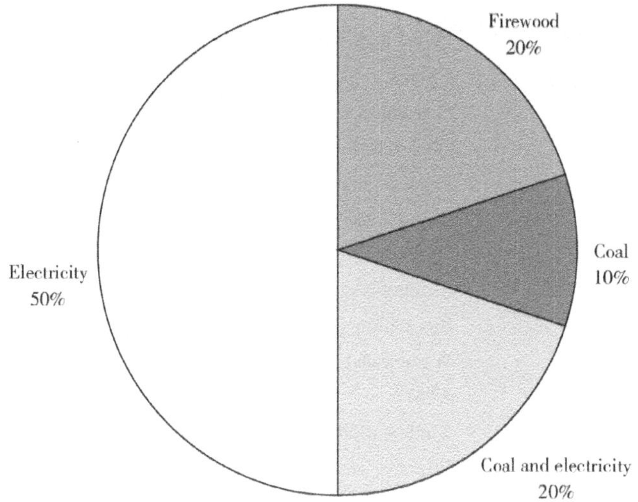

Fig. 2.1 Types of cooking fuel in migrant families

2.1.3 Public Facilities

After migration, the migrants have access to more public facilities in settlements and enjoy a variety of leisure activities. The ten villages all have essential public facilities, such as clinics and convenience stores, which have satisfied the residents' healthcare and daily needs. Among the ten villages, 90% have cultural activity centers where the residents participate in cultural activities, 70% have sports facilities to enable resident's sports activities, 60% have waste transfer stations or public trash cans which have improved hygiene and reduced pollution in the villages, and 50% have kindergartens or elementary and middle schools which help improve the education in these villages and the literacy of residents. Thirty percent of these villages have banks or ATMs to facilitate resident's daily transactions, 30% have bus stations facilitating resident's travel, 20% have farmers' markets where residents can buy a wide variety of goods, and 10% have department stores. So far, none of these villages has children's playground or similar facilities (See Table 2.2).

2.2 Economic Conditions of Ecological Migrants

This section describes the economic conditions of ecological migrants, including their household financial balances, household income, and individual income.

Table 2.2 Public facilities in settlements (%)

Public facilities	
Clinic	100
Convenience store	100
Department store	10
Farmers' market	20
Cultural activity center	90
Sports facility	70
Children's playground and similar facility	0
Bus station	30
Waste transfer station/public trash can	60
ATM	30
Kindergarten/elementary and middle school	50
Effective sample	10

2.2.1 Household Financial Balances

In 2015, the financial balance of migrant families in Ningxia averaged −20,448 yuan, showing that they were unable to make their ends meet. The income of migrant families was unable to cover their expense, leading to negative balances. Table 2.3 shows the household financial balances of migrant families in 2011 and 2015. In 2015, 67.54% of the migrant families were living on an inadequate income, 1.49% were able to maintain a balance of income and expenditure, and only 30% had a surplus. In 2011, the household financial balances of migrant families averaged 1631.62 yuan, and nearly 60% of the migrant families had a surplus (see Table 2.3). As can be seen, the financial situation of migrant families was worse in 2015 than in 2011.

Table 2.3 Household financial balances of migrant families (%)

Household financial balance (yuan)	2015	2011
10,001 and above	17.91	29.8
5001–10,000	5.22	11.9
1–5000	7.84	15.8
0	1.49	5.7
−1 to −5000	7.46	11.2
−5001 to −10,000	8.96	7.8
−10,001 and below	51.12	16.0
Effective sample (share)	80.4 (100.0)	76.9 (98.2)

Note Due to missing values in 2011, the percentages that year do not add up to 100%

2.2.2 *Household Income*

The household income of migrant families in Ningxia averaged 32,137 yuan in 2015. To be specific, families with household income of less than 10,000 yuan accounted for 15.4% of the total, and families with household income of 10,000–19,999 yuan, 20,000–29,999 yuan, and over 30,000 yuan accounted for 22.6%, 19.9%, and 42.0% of the total, respectively. In 2011, families with household income of less than 10,000 yuan accounted for 11.4% of the total, and those with household income of 10,000–19,999 yuan, 20,000–29,999 yuan, and over 30,000 yuan accounted for 30.9%, 23.5%, and 34.2% of the total, respectively. Generally speaking, the household income of migrant families has been increasing over the recent years. As for annual individual income, in 2015, people with an annual income of less than 1000 yuan, 1001–5000 yuan, 5001–9000 yuan, and over 9000 yuan accounted for 22.01, 20.02, 12.12, and 45.77% of the respondents. Compared with the per capita disposable income of permanent residents in rural areas of Ningxia of 9119 yuan,[1] the annual income of 54.23% of the respondents was significantly lower. In 2015, 29.35% of the migrants earned an annual income of less than 2980 yuan, who should be covered by precision-targeting poverty alleviation measures.

2.2.3 *Household Expenditure*

The migrant families spent 52,585 yuan on average in 2015. In 2011, the average household spending of migrants in general was 24,714.4 yuan. As a result of rising living cost, migrant families have seen significant increase in expenditure. Among the respondents, families with household spending of less than 10,000 yuan, 10,000–19,999 yuan, 20,000–29,999 yuan, and over 30,000 yuan account for 5.1%, 16.6%, 17.5%, and 60.8% of the total, respectively. Most of them spend more than their income.

2.3 Social Status of Ecological Migrants

The social status of people and their families is reflected in their status in the household registration system, education level, political status, resettlement, jobs, and others. This section analyzes the social status of the ecological migrants in Ningxia from the aforementioned perspectives.

[1] Statistics Bureau of Ningxia Hui Autonomous Region, "Statistical Communique of Ningxia Hui Autonomous Region on Ningxia's Economic and Social Development in 2015," May 25, 2015, http://www.cnstats.org/.

Table 2.4 Type of *hukou* of ecological migrants (%)

Item	Hukou status	Year of relocation	
		2011	2015
Type of hukou	Rural *hukou*	95.6	96.5
	Urban *hukou*	4.2	3.1
	Resident *hukou*	0.1	0.4
	Effective sample (share)	80.0 (100.0)	80.4 (100.0)

2.3.1 *Status in the Household Registration System*

Of all the ecological migrants in Ningxia, 96.5% are registered as rural residents (meaning that they have rural *hukou*), 3.1% are registered as urban residents (meaning that they have urban *hukou*), and only 0.4% have been registered as residents.[2] The proportion of ecological migrants with rural *hukou* increased from 95.6% in 2011 to 96.5% in 2015, while the proportion of people with urban *hukou* decreased (see Table 2.4). Thanks to progress in promoting the new type of urbanization, the urbanization rate in Ningxia has been rising during the past years and reached 55.23% in 2015. A major facet of the new type of urbanization is transformation of the status of rural residents in the household registration system. There have not been significant changes in the type of *hukou* of ecological migrants, which shows that the ecological relocation areas are falling behind in promoting new type of urbanization.

2.3.2 *Education Level*

The education level of residents has a significant influence on a region's productivity and economic development. The low income and poverty of ecological migrants are closely related to the overall low level of education. Among these migrants, 42.3% have not completed primary school, 22.3% have only completed primary school, and 23.4% have completed junior high school. Only 8.4% of the ecological migrants have been to vocational or general high school and technical school, and 3.6% have received associate degree or higher-level education. Over 40% are illiterate or have only completed primary school. Less than 10% have been to vocational or general high school and technical school, resulting in severe shortage of professional talent. A very small proportion of the migrants have been to junior college or have received higher-level education. Although the overall education level of

[2] China implements a dual household registration system under which people are registered as either rural or urban residents. In view of the changes in policies in the new era, such system can no longer satisfy the requirements of the time and reform is thus required. Pilot reform has been conducted in some cities in China to register all people as residents without specifying rural or urban residents.

Table 2.5 Education level of ecological migrants (%)

Item	Value	Year of relocation	
		2011	2015
Education level	Primary school and below	42.4	42.3
	Primary school	26.8	22.3
	Junior high school	21.7	23.4
	Vocational high school, general high school, technical school	7.4	8.4
	Junior college and above	1.7	3.6
	Effective sample (share)	79.9 (100.0)	80.4 (100.0)

these migrants remains low, compared with four years before, their education structure has been improving. As shown in Table 2.5, the number of people who have received professional training and higher education increased in 2015, showing that the migrants have been attaching importance to education as a way to fight poverty.

2.3.3 *Political Status*

There has not been much change in the political status of migrants since relocation. Most of the ecological migrants, or 92.3% of the total, are not members of the CPC or the Chinese Communist Youth League (CCYL). CPC and CCYL members account for 4% and 3.7% of the total, respectively.

The survey in 2011 showed that 90.1% of the migrants were not members of the CPC or the CCYL, 4.9% were CPC members, 4.9% were CCYL members, and 0.1% were members of non-communist parties (see Table 2.6). No significant changes are observed in the political status of ecological migrants in the past 4 years, which suggests that further efforts are required to improve the primary-level organizations in settlements.

Table 2.6 Political status of migrants (%)

Political status	2011	2015
No party affiliation	90.1	92.3
CCYL member	4.9	3.7
CPC member	4.9	4
Member of other parties besides CPC	0.1	0

2.3.4 *Means of Resettlement*

Most of the respondents have been relocated for some years. Of the total, 32.6% were relocated less than five years ago, 15.2% were relocated five to ten years ago, and 52.2% were relocated over ten years ago (see Fig. 2.2). In addition, the relocation has been conducted in a variety of ways. At the early stage of ecological migration, migrants were mainly relocated on a self-initiated basis and resettled in existing villages, and problems arose due to lack of plans. At a later stage, whole villages were relocated and merged to establish new villages, which has improved the situation. This approach allows the migrants to maintain their social relations and facilitates the planning, construction, and industrial development of settlements. Of all the respondents, 46.1% were relocated and resettled in whole-village relocation projects, 5.2% were relocated along with the whole village and resettled in existing villages in places of destination, 42.3% relocated on their own and resettled without government assistance, 5.9% were relocated on a self-initiated basis and resettled in existing villages in places of destination, and 0.5% were relocated and resettled by other means (see Fig. 2.3). As is obvious, a major proportion of the ecological migrants has relocated and resettled self-initiatedly without government assistance. Such self-initiated migration is found throughout all stages of ecological migration. Impressed with the remarkable achievements of ecological migration, the impoverished have, without being required to relocate, turned to their relatives or friends in settlements and settled down. Self-initiated migrants pose challenges to social governance in relocation areas, but they also contribute to the economic and social development of these areas. Their living conditions have also improved.

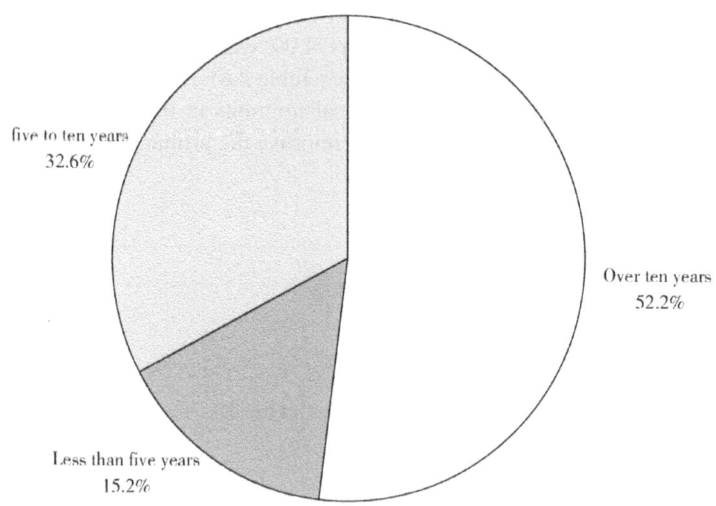

Fig. 2.2 Years since relocation of the migrants

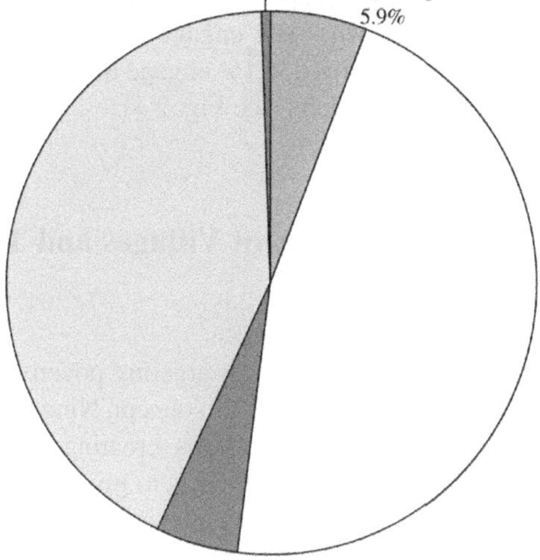

Fig. 2.3 Resettlement of migrants

Concerning the nature of the migration, 84.7% of the migrants said it was self-initiated, 13.8% claimed to be forced to move, and 1.5% could not specify the nature of their migration. Most of the migrants lived in mountainous areas with harsh natural conditions before the migration. The migration not only facilitates environmental protection in areas of origin, but also improves the lives of the migrants. Therefore, most people have responded to the call of the government and voluntarily moved to relocation areas. Data shows that 47.9% of the migrants have relocated self-initiatedly and 52.1% have been organized to relocate according to relevant policies. Government-organized migrants have access to preferential policies of the government of relocation areas in education, social security, housing, and land, but self-initiated migrants do not. Considering the large scale of self-initiated migration, in recent years, the government of relocation areas has included self-initiated migrants in their administration and introduced preferential policies for them in some areas. In particular, these migrants are given similar or the same preference as government-organized migrants in terms of children's schooling, social security, and other areas. As a result, the social governance in relocation areas has shown significant improvement.

2.3.5 *Jobs*

Ecological migrants have been doing a wide variety of jobs, and those living on farming alone do not constitute a big proportion. Of the total, 37.4% work in farming, 34.2% work outside their hometown, 3.8% engage in both farming and other jobs, 13.5% are unemployed, 0.7% are retired, 4.1% engage in finance, 0.9% are teachers, 2.4% are students, and 3% do other jobs (see Fig. 2.4).

2.4 Poverty of Ecological Migrant Villages and Their Plans for Poverty Elimination

Xi Jinping proposed the concept of precision-targeting poverty alleviation during a visit to Xiangxi, Hunan, in 2013. Guided by this concept, Ningxia has taken measures such as sending resident officials to poor villages, creating new business platforms for poverty alleviation, and introducing new ways to give financial support to fight poverty, shifting from indiscriminative poverty alleviation measures to targeted ones. Such transformation has caused significant changes in the life and production activities of the impoverished in Ningxia. This section analyzes the prominent problems in Ningxia's precision-targeting poverty alleviation of ecological migrants to enable its further progress.

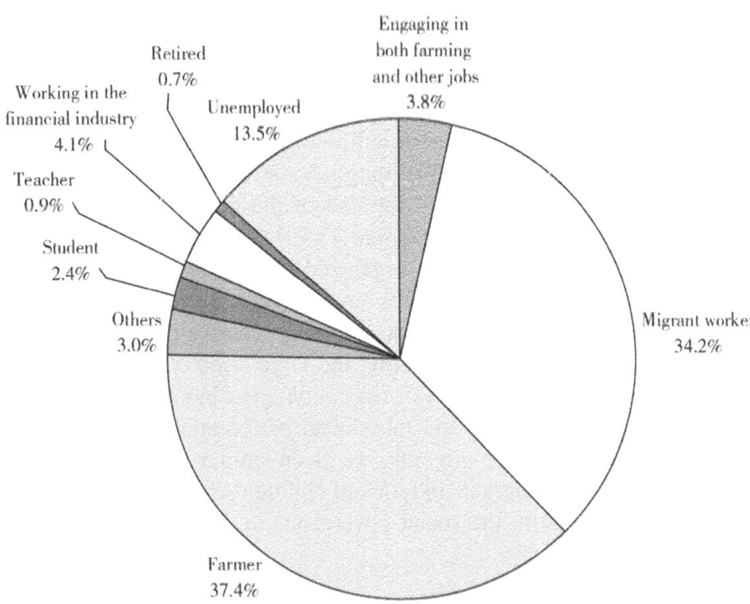

Fig. 2.4 Migrants' jobs

Up to 2015, there are approximately 580,000 people in Ningxia living in poverty, a considerable proportion being ecological migrants. The ecological migrants fall behind other Ningxia residents in economic development due to weaker economic foundation, inadequate resources, and late start. In 2015, the per capita disposable income of Ningxia's ecological migrants was 5830 yuan, far lower than that of Ningxia's rural residents, 9119 yuan, in the same year. It is evident that ecological migrants remain a weak link in Ningxia's efforts toward prosperity.

In the ten settlements covered in the survey, eight villages remain poverty stricken: Chaoyang, Dahe, Hongya, Lihua, Qihai, Tuanjie, Xinghai, and Yuanlong. Except for Yuanlong, Qihai, and Xinghai, the other settlements have been established for over five years. With greater efforts to alleviate poverty in recent years, resident poverty alleviation teams have been sent to each of the settlements, usually by county- or authority-level authorities. A few settlements have been paired with relevant government departments of Ningxia Hui autonomous region for poverty fighting. As further efforts are made to fight poverty in a targeted manner, all settlements have made poverty alleviation plans based on their respective conditions. According to their plans, Yuanlong, Mulan, Hongya, Lihua, Tuanjie, and Xinhua should cast off poverty in 2015, Chaoyang should shake off poverty in 2017, and Xinghai, Dahe, and Qihai should eliminate poverty in 2018. However, 80% of the officials of these villages thought it hard to eliminate poverty as planned, and only officials of two villages, Mulan and Xinhua, believed that they could complete the poverty elimination task as scheduled. Overall, most settlements in Ningxia have been confronted with onerous tasks, making it difficult for them to lift villagers out of poverty as planned. Village officials have set out several reasons, including unbalanced industrial structure, difficulty in finding jobs, inadequate water supply as a result of aging pumps, shortage of water resources, lack of good projects and development funds, large number of working-age people with disabilities and diseases, poor labor skills of migrants, and other problems. The reasons for each settlement are specified in Table 2.7.

The economic hardship of those living in settlements is measured in terms of the numbers of households receiving subsistence allowance and households registered as poverty stricken, as shown in Table 2.8.

Ningxia has been alleviating poverty in a targeted manner mainly by boosting business development, improving education, and social security. As shown in Table 2.8, in central Ningxia, Chaoyang, Dahe, Hongya, and Lihua of Wuzhong city and Xinghai and Qihai of Zhongwei city have a large group of ecological migrants that require precision-targeting poverty alleviation measures, and they are the main battlefields of precision-targeting poverty alleviation in Ningxia. In 2015, of 1420 households and 6397 people in Chaoyang, 2910 people in 653 households needed access to precision-targeting poverty alleviation measures, accounting for 45.99% of the total households and 45.49% of the total population. In Qihai, 2591 people in 492 households out of 887 households and 4349 people needed access to precision-targeting poverty alleviation measures, accounting for 55.47% of the total households and 59.58% of the total population. In Xinghai, 789 people in 312 households out of 612 households and 2962 people needed access to precision-targeting poverty alleviation measures, accounting for 50.98% of the total households and 26.64% of

Table 2.7 Major reasons causing difficulties in poverty alleviation in settlements

Village	Reason(s)
Chaoyang	Suboptimal industrial structure
Yuanlong	Land shortage Inability to work due to diseases or disabilities Difficulty in finding jobs
Xinghai	Land shortage Lack of funds Poor labor skills of migrants
Mulan	Lack of development funds
Dahe	Lack of good development projects Difficulty in finding jobs
Hongya	Aging pumps leading to inadequate water supply and difficulty in developing agriculture
Lihua	Lack of good development projects
Qihai	Lack of development funds and land
Tuanjie	Lack of development funds and land
Xinhua	Inability to work due to diseases and disabilities

the total population. Some settlements, especially those established during the 12th Five-year Plan period, still have a lot of work to do to fight poverty. settlements near Yinchuan city have found it relatively easier to complete the tasks of precision-targeting poverty alleviation. For example, there are only six people in one household in Mulan and around ten households, respectively, in Tuanjie and Xinhua requiring access to precision-targeting poverty alleviation measures.

Targeted poverty alleviation is an important approach to realizing shared prosperity. Its success is fundamental to the building of a prosperous Ningxia in all respects by 2018. Ecological migrants must seize the opportunities afforded by precision-targeting poverty alleviation programs to pursue development and improve their economic status.

2.5 Major Problems in Socioeconomic Development of Ecological Migrants

2.5.1 Shortage of Water and Farmland in Some Relocation Areas

As of 2015, Ningxia has relocated 1.16 million people, on sixth of its total population. Due to unlicensed farmland cultivation by some migrants, huge amount of irrigation water needed in arid areas, and expansion of crash crop, some relocation areas have

Table 2.8 Households receiving subsistence allowance and households registered as poverty stricken in settlements

Village	Households receiving subsistence allowance	Households registered as poverty stricken	Number of households having access to different poverty alleviation measures
Chaoyang	441 households 640 people	653 households 2910 people	Business development: 198 households, 891 people Financial measures: 158households, 679 people Education: 40 households, 208 people Skill training and labor export: 127 households, 536 people Social security: 130 households, 596 people
Yuanlong	35 households 43 people	73 households 351 people	Business development: 48 households Education: 13 households Social security: 12 households
Xinghai	265 households 487 people	312 households 789 people	Education: 74 households, 152 people Business development: 181 households, 524 people Social security: 57 households, 113 people
Mulan	67 households	1 household 6 people	Social security: 1 household, 6 people
Dahe	67 households	110 households	Business development: 78 households Social security: 15 households Education: 17 households
Hongya	387 people	206 households	Business development: 160 households Education: 25 households Social security: 21 households
Lihua	345 people	55 households	Education: 9 households Social security: 17 households Business development: 29 households

(continued)

Table 2.8 (continued)

Village	Households receiving subsistence allowance	Households registered as poverty stricken	Number of households having access to different poverty alleviation measures
Qihai	396 households 571 people	492 households 2591 people	Business development: 274 households Labor export: 1289 people Education: 610 people Social security: 418 people
Tuanjie	122 people	10 households	Social security: 8 households Education: 2 households
Xinhua	55 households 97 people	12 households	Social security: 12 households

been affected by shortage of water. According to Zhang Zhanguo, Party secretary of Hongya, a village in Dahe of Hongsipu district, the acutest problem for them is lack of water. "Hongsipu mainly uses water pumped from the Yellow River for irrigation. The project was designed to supply water for approximately 100,000 people. With 200,000 people now living in Hongsipu, each person has access to only half the amount of irrigation water they used to have. Crops cannot grow without sufficient water. The pump for the project, which was installed in 1998, does not have adequate capacity. If we do not work to address the issue of water shortage, farmers will all have to move out of this area."

As a flood of migrants have moved to different locations in Ningxia, the shortage of water and farmland in these areas is becoming increasingly acute. Moreover, the Yellow River is now in the dry season. As a result, Ningxia is experiencing water scarcity. The water deficit in Ningxia averages 10% and reached 23% in 2015. By 2020, there will be a shortage of 130 million m^3 of water in Ningxia. Population in the Yellow River pumping irrigation area, on the other hand, has been increasing dramatically. The migrant's demand for water has far exceeded the design capacity of the irrigation project at the early stage of migration. Meanwhile, as more and more people resettle in relocation areas, some of these areas have experienced shortage of farmland. In Dahe, a settlement established in 1998, each three-person household was given about 0.5 ha of farmland. In settlements established during the 12th Five-year Plan period, such as Xinghai and Qihai, on average, each migrant was given less than 0.07 ha of farmland. The shortage of farmland has had a severe negative impact on agricultural production.

2.5.2 Poverty of Elderly Migrants

An old Hui man surnamed Ma in Qihai, a village in Dazhanchang of Zhongning County, told us, "I receive 457 yuan every month, including 262 yuan in subsistence allowance, 145 yuan in old-age pension, and 50 yuan in rent income from leasing out my land. It's barely enough to feed myself with. My son and daughter-in-law are working in other places, and we take care of our grandson. They don't make much money as migrant workers, so they cannot support me. I cannot afford to get a cold, because it can cost me several hundred yuan. Before I was relocated from the mountain, I was able to grow more food than the family could consume. I also kept cattle and sheep, so I could earn 10,000–20,000 yuan. That's not the case now. Even the food I eat has to be brought down from where we lived before up in the mountain."

WXF, deputy Party secretary of Yuanlong in Minning town of Yongning County, is concerned about the living conditions of old people living alone in the village. He said, "There are a good number of old people living alone in our village. They cannot even prepare proper meals for themselves. The villager committee has designated people to send them meals. A main challenge is that we do not have enough money. We owed debts of over 100,000 yuan for taking care of these old people. We are applying for financial aid from civil affairs authorities, but it is very hard."

The difficult situations of the old man surnamed Ma are a microcosm of the hardship of some elderly migrants. Old migrants face four main problems. First, it has been hard for elderly migrants to earn income from farming or animal husbandry due to limited and poor land resources in relocation areas. Second, elderly migrants who have lost work ability are unable to earn any income, and they can only maintain a low living standard relying on subsistence allowance or basic old-age pension. Third, as a result of the weakening concept of filial piety and the handover of responsibility toward family, the traditional lifestyle of offspring caring for the elderly is being replaced by the elderly caring for their offspring. Instead of receiving support from their children in life and finances, senior people need to take care of their grandchildren and other family members. Fourth, influenced by the deep-rooted concept that offspring shall care for their elderly parents, the bulk of migrants prefer home-based elderly care. Old migrants living alone who cannot care for themselves have to turn to villager committees for help. Due to inadequate financial resources, however, it is hard for villages to provide them with constant aid. It is also hard to promote community-based elderly care and private elderly care by non-government agencies in relocation areas.

2.5.3 Lack of Further Impetus for Industrial Development

Thanks to a succession of good harvests of crops, the prices of agricultural produce have been declining, resulting in decreasing income of migrant families whose income mainly comes from crop production. The increase in costs of agricultural supplies further reduces the profit of crop production. The profit of cash crops such as goji berries and grapes has also dropped.

According to SXC, Party secretary of Dahe village in Hongsipu district of Wuzhong city, "the large-scale planting and processing of goji berries by large enterprises have made it difficult for individual households planting goji berries to make a profit. The size of goji berry fields dropped from 140 ha in 2013 to approximately 21 ha in 2015. Most villagers have started growing corn or clover instead of goji berries. Some have gone to work in enterprises."

People relocated in the 12th Five-year Plan period have been granted fewer farmland the quality of which is poor. The farmland in some relocation areas is not arable due to desertification or salinization.

MJP, Party secretary of Xinghai village in Xuanhe town of Zhongwei city, told us, "we were resettled in Xinghai in 2011, but did not receive our farmland until 2015. Each villager is given less than 0.07 ha of farmland, and the bulk of the land has been affected by desertification. No plants can grow. It's too costly. We have reached an land transfer agreement with an enterprise at the price of about 5700 yuan per hectare per year. We cannot make a profit from our farmland."

Most migrants intend to raise livestock in their courtyards, but their courtyards are too small. In addition, as most settlements established in the 12th Five-year Plan period have not created a cooperation mechanism comprising the government, enterprises, and rural households in development of animal husbandry. The interests of government, enterprises, and rural households are not balanced.

As a result of the decline of construction industry and other labor-intensive industries, migrants have seen a significant decrease in their work income. In addition, due to limited number of industrial parks near settlements, the migrant's income from working in enterprises is rather unstable. It is even harder for migrants aged 50–60 to find jobs and make income, as they are inferior to young people in physical condition, health, and skills.

A migrant surnamed Li in Qihai village told us, "Previously I could earn about two hundred yuan each day as a mason, and it was easy to find jobs. This year is harder. My wage is also lower. I can make only seventy to eighty yuan a day."

Industrial development has an immediate impact on migrant's income. The migrant's businesses are currently at a primary stage of extensive development, with a homogeneous industrial structure. This has affected the stability and sustainability of migrant's income from industrial development. Migrants resettled in the 12th Five-year Plan period have found it hard to develop plant production and animal husbandry due to limited resources. Employment service has thus become a pillar industry in settlements. The waning real economy, however, has undermined the stability of labor service and led to significant decrease in income of some migrant families. In addition, after relocation, the migrant's expenditures, especially consumption expenditure, have increased. A lot of migrant families have become poverty stricken because their income falls short of their expenditures. Another major limitation on further business development of the migrants is inadequate funds. With limited capital, the migrants often have to borrow money from their relatives and friends to start businesses. They show a strong intention of applying for loans, but it has been hard for them to obtain loans. In short, without multiple sources of funding, the migrants cannot start or further develop their businesses. Consequently, we have not seen significant increase in migrant's income, and the progress in poverty alleviation has been slow.

2.5.4 Inaccurate Identification of the Impoverished in Some Villages

Targeted poverty alleviation attaches great importance to precise identification of the poor which is the precondition for all poverty alleviation work. In Ningxia, at present, poverty identification is conducted with county and township, instead of village, as a basic unit. Moreover, disparities exist between villages in criterion of poverty measurement. Strict procedures are established to ensure the precision of poverty identification, according to which eligible households are registered as poverty stricken after submitting applications, democratic appraisal, preliminary examination and publishment by village authorities, re-examination and publishment by township authorities, and approval and announcement by authorities at the city, county, or district level. In practice, however, most of the poor migrants are not familiar with precision-targeting poverty alleviation policies and rarely take the initiative to submit applications. The relevant households are not appraised in a democratic manner, and problems are found in examination at various levels. The households to be registered as poverty stricken and receive assistance are usually determined by officials of the villager committee and township government, some of which are reported to higher-level authorities for approval without proper publishment. All these have led to negative comments on and even dissatisfaction with poverty identification procedures among the migrants. In addition, the poverty line in Ningxia is 2980 yuan per person per year. In poverty identification, the poverty is measured based on household's income, without considering their expenditures. It is rarely based on comprehensive measurement of the conditions for migrant families' lives and production activities, development capacity, human resources, and stability of income. Some migrant families which have experienced a dramatic increase in expenditures due to diseases or education expense have not been covered by precision-targeting poverty alleviation policies because their per capita disposable income is above the poverty line.

A villager surnamed Wang in Chaoyang village of Hongsipu district said, "I have been working outside the village. I had no knowledge of precision-targeting poverty alleviation in the village until the poverty identification results are published. I have a family member of poor health. We spend a lot on medicine and medical treatment, and we are burdened with heavy debts. We have a hard life, but we have not been identified as a poor household because we cannot meet the criteria. It's not fair."

Chapter 3
Production and Living Conditions of Migrants

Lei Fan

In Ningxia's ecological migration project, it is of equal importance to improve the ecosystem in places of origin and improve the living conditions of the migrants. In promoting the project, ecological protection is integrated with development-oriented poverty alleviation to boost local economic and social development and improve the people's living standard. Due to harsh natural environment and scarcity of natural resources, the migrants, before relocation, lived in poverty for a long time. Their over-exploitation of limited natural resources for survival further damaged the ecosystem. Therefore, while the ecological migration project is supposed to restore and improve the ecology, people also hope that the migration can help overcome the difficulties in their life and production activities. In particular, in the ecological migration project, development-oriented poverty alleviation measures are adopted instead of merely providing relief. Policy support, financial support, and technical support are given to migrants at different stages of their production activities, including supply, production, and sale, to enhance their abilities to improve and develop themselves. With development of local economy and society, multiple approaches to alleviating poverty among the migrants have been adopted, such as promoting industrial development, entrepreneurship, and labor export, which has improved the migrant's conditions of life and production activities. The development of characteristic agriculture and industrial parks in ecological migration in Ningxia has significantly increased the channels for the migrants to find employment, laying a foundation for improving the household income and living standards of the migrants.

L. Fan (✉)
Institute of Sociology, Chinese Academy of Social Sciences, Beijing, China
e-mail: fan-lei@cass.org.cn

X. Wang (ed.), *Ecological Migration and Targeted Poverty Alleviation in Ningxia*, International Research on Poverty Reduction,
https://doi.org/10.1007/978-981-19-7888-3_3

3.1 Income and Expenditure of Migrant Families

The ecological migration project has raised the living standards of migrant families. First, the majority of migrant families have seen an increase in their income, and their income from nonfarm activities has been growing faster, showing that development-oriented poverty alleviation measures have been effective. Second, the consumption expenditure of migrant families has increased, while the percentage of spending on food has decreased. Nonetheless, the migrants are still in the lower-middle group in Ningxia in terms of per capita income, and a large proportion of migrant families live in extreme poverty or rely on subsistence allowance. Spending on housing and household appliances as a result of the migration makes up a large percentage of migrant's household expenditure. The household balances of migrant families are worrisome, and a large proportion of them are in debt. It is vital to change the migrant's consumption mindset and behavior and improve their consumption structure in order to improve their lives.

3.1.1 Household Income

According to our survey, household income and per capita income of migrants have both increased after the ecological migration. After migration, 61.2% of the respondents reported an increase in household income, 27.2% reported a decrease, and 11.6% reported no change in income. It is evident that the ecological migration project has played a positive role in boosting migrant's income. Meanwhile, the per capita income of migrants has also increased, growing from 5249 yuan[1] in 2012 to 6260 yuan in 2016, registering a nominal growth of 19.3%. The growth shows that the ecological migrants have enabled the migrant families to develop themselves so that they can maintain income growth in tandem with local economic and social development.

The income structure of migrant families shows three distinctive characteristics in 2016. First, the percentage of wages continued to increase. The per capita wage of migrants increased from 2743 yuan in 2012 to 3621 yuan in 2016, and the percentage of which in per capita income rose from 52.3 to 57.8%. In the meanwhile, the percentage of household income from agricultural activities showed a significant decrease. The per capita income from agricultural activities of migrants in 2012 stood at 2269 yuan, accounting for 43.2% of the per capita income. In 2016, it dropped to 1562 yuan, accounting for 24.9% of the per capita income. The data shows that wages have become the most important source of income for migrant families. As wages have been growing faster than per capita income, they have become the main factor enabling a higher living standard. Second, the percentage

[1] The data are cited from the 2012 survey of ecological migration in Ningxia. The income and expenditure data in this chapter reflect the situation in 2011 and 2015, but were collected in 2012 and 2016, and thus referred to as data in 2012 and 2016.

of income from business operation increased. In 2012, the per capita income of migrants from business operation was 91 yuan, accounting for 1.7% of the per capita income. In 2016, it increased to 318 yuan, accounting for 5.1% of the per capita income. The increase shows that migrant families have acquired better abilities to start businesses and that more households are engaging in business operation. The percentage of migrant families which have started businesses increased from 1.3% in 2012 to 4.1% in 2016. Meanwhile, their businesses have been expanding. The highest income among migrants from business operation in 2012 was 80,000 yuan, which rose to 500,000 yuan in 2016. Third, the percentage of transfer income, including income received in the form of social security, social aid, government subsidies, and benefits paid by villages, showed a significant increase. The per capita transfer income of migrants increased from 119 yuan in 2012 to 674 yuan in 2016, and its percentage in the total per capita income was up from 2.3 to 10.7%. Apparently, thanks to government's greater efforts to meet people's basic living needs, especially the expansion of the social security system, the share of transfer income in total income of migrant families has experienced a significant increase (see Table 3.1).

Similar characteristics are displayed in the main sources of income before and after the migration, as reported by the respondents. First, the percentage of income from plant production in total household income experienced a sharp decline, dropping from 86.4% before the migration to 27.2%, while the percentage of wages

Table 3.1 Income structure of migrant families

Source of income	2012		2016	
	Per capita income	Share	Per capita income	Share
Agricultural income	2269	43.2	1562	24.9
Wages	2743	52.3	3621	57.8
Income from business operation	91	1.7	318	5.1
Property income	27	0.5	85	1.3
Transfer income	119	2.3	674	10.7
Total	5249	100.0	6260	100.0

Unit Yuan, %

Note In this research, agricultural income refers to household income derived from various agricultural activities. Wages include wages, bonuses, allowances, and goods received on holidays. Income from business operation includes all income from nonfarm business operation. Property income includes house rental income, income from transfer of land right, and income from financial investment such as bonds, deposits, loans, securities, dividends, and so forth. Transfer income includes income from social security such as pensions, old-age insurance, unemployment insurance, work-related injury insurance, and maternity insurance of family members, reimbursement of medical expense, financial aid provided by government, employer, and non-government organizations, such as subsistence allowance, allowance for people in need, medical aid, disaster relief, school scholarship, financial aid for poor students, government subsidies to businesses, income derived from supportive policies such as agricultural subsidies, tax reduction and exemption, and benefits provided by neighborhood/villager committees such as dividends of collective economy and subsidies

Table 3.2 Main sources of income of migrant families before and after the migration

	Main source of household income before migration		Main source of household income after migration	
	Number of people	Share	Number of people	Share
Plant production	695	86.4	219	27.2
Animal husbandry	33	4.1	31	3.9
Working for others	72	9.0	495	61.6
Business operation	1	0.1	20	2.5
Rental	0	0.0	1	0.1
Wages as employee of administrative and public organizations	2	0.2	4	0.5
Subsidies and financial relief	1	0.1	13	1.6
Others	0	0.0	21	2.6
Total	804	1000	804	1000

Unit Person, %

in household income increased from 9.0% before the migration to 61.6%. Of all the respondents, in 2016, 62.2% claimed wages as the new and biggest source of income. The percentage was 53.3% in 2012. Second, the percentage of households with business operation as the main source of income increased from 0.1% before the migration to 2.5%. Third, the percentage of households with subsidies and financial relief as the main source of income increased from 0.1 to 1.6% (see Table 3.2).

Nonetheless, we have observed a low level of satisfaction with income among migrants. The survey shows that 24.0% of the respondents were very dissatisfied with their income in 2015, 38.3% were not satisfied, 20.3% were neutral, 16.7% were satisfied, and 0.7% were very satisfied (see Table 3.3). The ecological migration project has improved the income structure of migrant families and raised their income, but compared with the average household income of Ningxia, the migration families are still in the lower-middle group. According to the Statistical Communique of Ningxia Hui Autonomous Region on Ningxia's Economic and Social Development in 2015, the per capita disposable income of rural residents in Ningxia stood at 9119 yuan. The per capita income of the migrants accounts for less than 70% of the average figure of Ningxia. The ecological migration project has satisfied the migrant's basic living needs and laid a foundation for their further development. But problems such as huge consumption expenditure, high cost of agricultural supplies, and instability of nonfarm employment have a grave negative impact on migrant's satisfaction with their income.

Table 3.3 Are you satisfied with your income in 2015?

	Number of people	Share
Very dissatisfied	193	24.0
Not satisfied	308	38.3
Neutral	163	20.3
Satisfied	134	16.7
Very satisfied	6	0.7
Total	804	100.0

Unit Person, %

3.1.2 *Household Expenditure*

The per capita household expenditure of migrants has also increased after the migration, as shown in the figures in 2012 and 2016.

According to the survey in 2016, after the migration, 95.1% of the respondents experienced an increase in their household expenditure, 2.1% saw a decrease, and 2.7% experienced no changes. The main items resulting in the increase in household expenditure include daily expenditure (47.1%), medical care (21.8%), education (11.2%), plant cost (10.6%), gift money for weddings and funerals (5.3%), house building (2.0%), commuting and communication expense (1.0%), cost of animal husbandry (0.8%), and others (0.3%). It is evident that the rising cost of living and production activities is the main factor leading to the increase in household expenditure of migrants.

Comparing the structure of household expenditure of migrant families in 2012 and 2016, we note the following characteristics:

First, the percentage of expenditure on food, clothing, transportation, communication, and entertainment has decreased. While the per capita food expenditure of migrant families increased from 1119 yuan in 2012 to 1674 yuan in 2016, its percentage in the per capita household expenditure dropped from 22.4% to 15.4%. The clothing expenditure, 281 yuan in 2012 and 371 yuan in 2016, experienced a decrease in its percentage in the per capita household expenditure from 5.6 to 3.4%. The spending on transportation and communication, 307 yuan 2012 and 538 yuan in 2016, was down from 6.1 to 5.0%. The percentage of spending on cultural and entertainment goods and services also decreased from 11.2% in 2012 to 7.0% in 2016. It is our preliminary conclusion that in view of the rising consumption expenditure and greater percentage of other expenses, the migrants have been cutting their expenditure on food, clothing, and transportation, which has resulted in the decrease in the percentage of expenditure on daily life.

Second, the percentage of expenditure on housing and household appliances and services has increased. As expenditure on house building, rebuilding, upholstery, and purchase of household appliances is essential for the resettlement, a major characteristic in the migration is that the percentage of expenditure on housing and household appliances and services has been increasing. Data shows that the increase in housing expenditure is not unique to new migrants. The migrant families relocated in 1993 have also made considerable expenditure on housing. Due to overcapacity, the prices of China's building materials hit a record low in the 12th Five-year Plan period in 2015. As a result, some rural residents increased their spending on house building and rebuilding, leading to an increase in household housing expenditure. The survey in 2012 showed that 8% of the respondents had made expenditure for house building and rebuilding. In 2012, the percentage rose to 14.7%.

Third, the percentage of spending on weddings and funerals has shown a significant increase. In 2012, 10.5% of the respondents had expenditure on weddings and funerals, and such expenditure accounted for 1.5% of their household expenditure. In 2016, 45.1% of the respondents had such expenditure which accounted for 14.4% of their household expenditure. Considering that 15.9% of the respondents spent 3000 yuan and less on weddings and funerals, it is estimated that some respondents included gift money which should be considered gift expenditure in spending on funerals and weddings. Overall, according to the survey in 2016, the expenditure on housing, weddings and funerals, and gifts account for a considerable proportion of migrant's household expenditure, which has resulted in the decrease in expenditure on daily life (see Table 3.4).

According to the Statistical Communique of Ningxia Hui Autonomous Region on Ningxia's Economic and Social Development in 2015, the per capita consumption expenditure of rural residents in Ningxia in 2015 was 8415 yuan. For migrant families, the per capita consumption expenditure was 10,870 yuan. The huge consumption expenditure has become a heavy burden on migrant families. The excessive expenditure on gifts has been a long-standing problem among migrants, which has to some extent undermined their efforts to improve their living standard. Meanwhile, statistics show that the growth rate of per capita disposable income of rural residents, which dropped from 14.2% in 2012 to 8% in 2016, has been decreasing for four years in a row since 2012. Chances are small that the migrant families in Ningxia may experience a significant increase in their per capita income. As the per capita consumption expenditure of migrants has been surging, migrant families are expected to face major difficulties in improving their lives. It is vital to change the migrant's consumption mindset and behavior and improve their consumption structure in order to improve their lives. Participation in diverse cultural activities and changing the way of networking will also improve the quality of their lives.

3.1.3 Structure of Income and Expenditure of Different Types of Rural Households

The ecological migration project has accelerated the progress of poverty alleviation among the migrants. Currently, migrant households living in extreme poverty and relying on subsistence allowance account for 25.0 and 23.8% of the total, respectively, showing that nearly half of the migrant households are still living in poverty.

Table 3.4 Structure of migrant's household expenditure on daily life

Item	2012		2016	
	Per capita expenditure	Share	Per capita expenditure	Share
Food	1119	22.4	1674	15.4
Clothing	281	5.6	371	3.4
Housing	456	9.1	1400	12.9
Household goods and services	484	9.7	1306	12.0
Transportation and communication	307	6.1	538	5.0
Cultural and entertainment goods and services	558	11.2	765	7.0
Medical care	912	18.2	2018	18.6
Supporting family member(s) not living together	39	0.8	149	1.4
Weddings and funerals	75	1.5	1565	14.4
Gifts	633	12.7	948	8.7
Others	134	2.7	136	1.3
Total expenditure	4998	100.0	10,870	100.0

Unit Yuan, %
Note Food expenditure includes spending on home-cooked food and dining out (prices of self-produced food are estimated and included in the spending). Housing expenditure includes rental paid, down payment made in 2011, monthly mortgage payments, and spending on home rebuilding and upholstery. Expenditure on household goods and services includes spending on household appliances, furniture, vehicles, power, water, gas (coal), and property management fee and heating fee. Expenditure on cultural and entertainment goods and services includes education expense, such as tuition, miscellaneous fee, stationary fee, spending on after-school classes, and boarding fee

Meanwhile, rural households enjoying "five guarantees" (i.e., guarantees of food, clothing, medical care, housing, and burial expenses), average rural households and rich rural households account for 1.2%, 48.7%, and 1.0% of the total, respectively, meaning that more than half of the rural households have been lifted out of poverty. Among the early migrant families relocated in the 1980s, approximately 30% are living below the poverty line, nearly 70% have been lifted out of poverty, and 3.6% have become rich (see Table 3.5). The percentage of poor households in recently relocated migrant families is still high. More than 50% of the households relocated since the late 1990s are still impoverished. As the specific type of household is determined by the respondents according to their own understanding, the results may not be the same as government statistics or evaluation results. Migrant households that have been registered as poverty stricken account for 13.6% of the total. The per capita income of these households is 4998 yuan, lower than the per capita income of 6260 yuan of all migrants. Data shows that the *Diaozhuang* project in the 1980s

Table 3.5 Share of different types of rural households based on respondent's self-assessment

	Before 1982	1983–1997 Diaozhuang project	1998–2000 "1236" project	Since 2000 Ecological migration	Total
Extremely poor	0.0	18.8	24.1	26.7	25.0
Enjoy subsistence allowance	50.0	12.5	28.5	24.9	23.8
Enjoy "five guarantees"	0.0	0.0	0.7	1.6	1.2
Average	50.0	65.2	45.3	46.1	48.7
Rich	0.0	3.6	0.7	0.5	1.0
Others	0.0	0.0	0.7	0.2	0.2
Total	100.0	100.0	100.0	100.0	100.0

Unit %

enabled a decrease in poverty ratio of migrant households from over 70 to 15.7%, and the "1236" project in the 1990s enabled a decrease in poverty ratio of migrant households from 69.9 to 13.9%.[2] The poverty ratio in Ningxia in 2016 is 14.5%.[3] In view of these data, the percentage of registered poverty-stricken households is consistent with the official statistics of poverty-stricken households.

The per capita income of self-proclaimed extremely poor households is 4484 yuan, lower than that of registered poverty-stricken households. For households receiving subsistence allowance and the "five guarantees," the per capita income is, respectively, 5619 yuan and 6265 yuan, close to that of all migrants. Average and rich rural households enjoy higher per capita income (see Table 3.6). Based on the self-proclaimed household income of migrants, there may be more poverty-stricken households than officially registered ones. Rich families derive one third of their household income, respectively, from agricultural activities, wages, and business operation, showing greater stability of income than other types of families. Other families rely on wages as a major source of income. Transfer income constitutes a large proportion of the income of extremely poor households and households receiving subsistence allowance and the "five guarantees," showing that government aid has played a significant role in maintaining the stability of the household income of these families. But transfer income may not effectively alleviate the poverty of extremely poor households, as it makes up a smaller percentage of the total household income of these families by comparison with those receiving subsistence allowance and the "five guarantees." Therefore, in addition to giving full play to

[2] CPC History Research Office of Ningxia Committee of the CPC, *30 years of rural reform and development in Ningxia* (Ningxia: Ningxia People's Publishing house, 2010).

[3] China Banking Regulatory Commission Ningxia Bureau and China Insurance Regulatory Commission Ningxia Bureau, "Ningxia's plan for precision-targeting poverty alleviation by characteristic industries," accessed September 19, 2016, http://www.nx.gov.cn/zwgk/gtwj/bmwj/133101.htm.

the role of the ecological migration project in development-oriented poverty alleviation, authorities must ensure precise identification of the impoverished households, take targeted measures to help them, perform precise management, and continue to provide financial aid to poverty-stricken migrant households.

In terms of the percentage of gift spending in per capita household expenditure, households enjoying the "five guarantees" rank first, although their per capita household expenditure is the lowest. Extremely poor households, households receiving subsistence allowance, and average rural households do not show much difference in per capita expenditure. The extremely poor households and households receiving subsistence allowance make a larger proportion of expenditure on medical care than other households, showing that illnesses-related poverty is a major cause of relative poverty. Expenditure on funerals and weddings also constitutes a large proportion of their household expenditure. A large proportion, 41.4%, of the household expenditure of rich families is for housing (see Table 3.7). The survey shows that besides medical care spending, the expenditure on funerals and weddings, gifts, and housing constitutes a large proportion of the total. Rural households, to varying degrees, have engaged in "face consumption," i.e., consumption not to satisfy practical needs but

Table 3.6 Income structure of different types of rural households

	Extremely poor households	Households enjoying subsistence allowance	Households enjoying five guarantees	Average households	Rich households	Others
Per capita income	4484	5619	6265	8164	33,522	6429
Share of per capita income from agricultural activities	25.2	24.2	35.7	24.9	30.9	19.4
Share of per capita income from wages	61.0	51.5	50.1	60.3	30.0	80.6
Share of per capita income from business operation	0.1	2.6	0.0	4.1	32.9	0.0
Share of per capita property income	1.7	2.0	0.3	0.9	0.0	0.0
Share of per capita transfer income	12.0	19.7	14.0	9.9	6.2	0.0

Unit Yuan, %

to flatter one's vanity, so as to keep up with each other. So far, governments at all levels have paid great attention to factors causing poverty such as illness and excessive education spending and have taken effective measures to address these issues. But it has been hard to eradicate "face consumption" in rural areas, despite repeated instructions from the government. In a sense, the increase in nonfarm income of rural households has widened the income gap between households and given rise to more hidden income. It has been common in villages that people like to keep up with the Joneses and show off their nonfarm income. So, in advancing the ecological migration project, in addition to lifting the poor out of poverty, efforts are also required to transform the customs in relocation areas. Otherwise, the effect of the migration project on poverty alleviation will be significantly undermined. Since the survey in 2012, the migrant's spending on funerals, weddings, and gifts have remained high. To some extent, "face consumption" has become a social issue with negative impact on the effect of poverty alleviation measures.

3.1.4 Household Debt

As the migrants are in the lower-middle group in Ningxia in terms of per capita income, but their per capita expenditure is more than the average in Ningxia, lots of migrant families are in debt. In the survey, 70.3% of the respondents said that they were in debt. Households in debt accounted for 85.6% of extremely poor households, 7.8% of households receiving subsistence allowance, 70% of households enjoying the "five guarantees," 58.8% of average households, and 50% of rich households.

The top three causes of their household debt are expenditure exceeding income (42.6%), illness (38.5%), and education expense (18.7%). Of the "other causes" that account for 28% of the total, 7.7% is related to home purchase, building, and renovation, 4.4% is related to investment, 3.7% is related to weddings, and 1.5% is related to gifts spending (see Table 3.8). Nearly 60% of the household debt has been caused by the migration. While the ecological migration has boosted the household income of migrants, it has also resulted in rising expenditure for these people, which has been underestimated. Before the migration, people living in harsh environment had to suppress their needs in lives and production activities, leading to both low income and low expenditure. While the migration has enabled them to make more income, the migrants have also experienced an increase in household expenditure due to growing need in lives and production activities, fine management of production, diversifying operations of households, and other needs. The increase in migrant's income is not sufficient to cover their growing expenses. As a result, a large proportion of the migrants have been in debt. Similarly, before the migration, due to limited access to medical service, people often did not seek medical treatment even if they were not well. After the migration, the migrants have access to better medical service so they can enjoy timely treatment, leading to an increase in medical expenditure and debt due to illness. Therefore, instead of only focusing on raising people's income in the ecological migration project, we should also pay attention to the increase

Table 3.7 Expenditure structure of different types of rural households

	Extremely poor households	Households enjoying subsistence allowance	Households enjoying five guarantees	Average households	Rich households	Others
Per capita expenditure	10,464	11,582	8601	10,968	31,733	14,107
Share of per capita food expenditure	16.5	15.3	15.7	16.8	5.7	25.3
Share of per capita clothing expenditure	2.5	3.4	3.1	4.2	4.3	7.0
Share of per capita housing expenditure	11.4	18.2	3.8	23.6	41.4	0.0
Share of per capita expenditure on transportation and communication	3.5	5.0	3.7	6.5	7.7	5.1
Share of per capita expenditure on cultural and entertainment goods and services	9.2	6.2	8.9	6.4	15.5	27.8
Share of per capita expenditure on medical care	28.8	21.5	17.3	15.1	17.7	1.5
Share of per capita expenditure on supporting family members not living together	2.1	0.7	1.5	1.4	0.3	3.5
Share of per capita expenditure on funerals and weddings	14.5	21.1	15.4	13.3	1.0	26.6
Share of per capita expenditure on gifts	8.5	8.6	15.2	10.2	6.5	3.2

(continued)

Table 3.7 (continued)

	Extremely poor households	Households enjoying subsistence allowance	Households enjoying five guarantees	Average households	Rich households	Others
Share of per capita expenditure on other items	3.1	0.0	15.5	2.5	0.0	0.0

Unit Yuan, %

in migrant's household expenditure and see if such increase has put a burden on them while counteracting the increase in income. The ecological migration project is the most convenient approach to ecological protection, which has also ensured the migrant's income by means of resettlement and supportive policies. Nonetheless, it has limited effect in balancing the migrant's expenditure and optimizing the expenditure structure. The survey shows that migrant's consumption behavior has to a great extent delayed the progress of poverty alleviation.

Table 3.8 Causes of household debt

	Extremely poor households	Households enjoying subsistence allowance	Households enjoying five guarantees	Average households	Rich households	Others	Total
Expenditure exceeds income after migration	42.1	45.3	28.6	42.1	50.0	0.0	42.6
Illness	42.7	41.9	28.6	34.2	0.0	0.0	38.5
Education expense	19.9	19.6	42.9	16.7	0.0	50.0	18.7
Migration expense	12.3	8.8	14.3	5.7	0.0	50.0	8.7
Loss of working ability due to aging	4.7	7.4	0.0	3.9	0.0	0.0	5.0
Disability	0.6	6.8	0.0	1.3	0.0	0.0	2.5
Others	25.1	20.3	28.6	35.1	50.0	0.0	28.0

Unit %
Note This is a multiple-choice question

3.2 Production of Migrant Families

In production activities of migrant households, nonfarm employment is taking an increasingly important position, while agricultural activities have been relegated to a secondary role. Rural households have increased their spending on fertilizers and water to adapt to the conditions for agricultural activities in relocation areas, but agricultural development has been primarily constrained by lack of funds and farmland. A large proportion of the migrant families are working in other places as migrant workers, and the income derived thereof accounts for a big share of their total income. But the migrants have been facing difficulties finding nonfarm employment due to problems such as instability of income as migrant workers, limited channels to find employment, inadequate access to job information, small chances of getting a job, inadequate ability of nearby industrial parks to create jobs for migrants, and so forth.

3.2.1 Agricultural Production

After the migration, dramatic changes have taken place in the way the migrants perform household operations. Agricultural income is no longer the biggest source of income. In addition, 50% of the migrant households have stopped engaging in agricultural activities. In terms of the income structure, 13.4% of the migrant families are only engaged in agricultural activities, 46% are only engaged in nonfarm businesses or jobs, 37.1% are engaged in both agricultural and nonfarm activities, and 3.5% live on property income or transfer income.

In the *Diaozhuang* project, 70.8% of the migrants were engaged in both agricultural and nonfarm activities. A large proportion of households relocated in the "1236" project were engaged in both agricultural and nonfarm activities, and the households only doing agricultural activities accounted for 25.5% of the total. Lots of households relocated during the ecological migration project are only engaged in nonfarm activities, accounting for 58.6% of the total (see Table 3.9). It is evident that thanks to the growing number of nonfarm jobs over the recent years, the percentage of people turning to nonfarm jobs or businesses instead of agricultural production has been on the rise.

Regarding the investment in agriculture, 84.1% of the households with agricultural business have increased their investment in agricultural production, 9% have reduced the investment, 5% have neither increased nor reduced the investment, and 2% are not sure about the change. According to the respondents, the increase in investment in agricultural production mainly goes to fertilizer (92%), irrigation (84%), equipment leasing (17.5%), hiring people (10.9%), and so forth. Compared with the areas where the migrants originally lived, the relocation areas enjoy better farmland,

Table 3.9 Production activities of migrant families at different times

	Before 1982	1983–1997 Diaozhuang project	1998–2000 1236 project	Since 2001 Ecological migration
Agricultural activities	0.0	8.8	25.5	11.4
Agricultural and nonfarm activities	50.0	70.8	51.8	26.3
Nonfarm activities	0.0	19.5	18.2	58.6
Others	50.0	0.9	4.4	3.6
Total	100.0	100.0	100.0	100.0

Unit %

adequate water, favorable industrial policies, advanced agricultural techniques, well-developed agricultural market, and other advantages. On the other hand, more investment in agriculture and advanced techniques are required. Therefore, the farmers have increased their expenditure on fertilizers and irrigation to satisfy the requirement for agricultural production in relocation areas.

In terms of the funding for agricultural production, self-funding accounts for 72.5% of the total, and loans and money borrowed from relatives and friends account for, respectively, 46.9% and 37.3%. Households only engaging in agriculture have a greater demand for funds in this regard. While 80.4% of these households are funded by their own money, 62.6% also get loans to make more investment in agricultural operation.

To varying degrees, the development of household plant production has been hindered by inadequate financial input (53.3%), shortage of farmland (42.3%), shortage of water (35.1%), inadequate agricultural equipment (17.5%), poor quality of farmland (15%), lack of planting techniques (11.1%), limited sales channels (4.4%), lack of good varieties (3.9%), and other problems.

As for household animal husbandry, its development has been hindered by inadequate financial input (54.5%), lack of space (33.5%), lack of techniques (18.2%), inadequate equipment (13.3%), lack of good breeds (7.2%), limited sales channels (4.8%), and other problems. The traditional way of animal breeding is still prevalent, as 83.4% of the respondents intend to build pens in their courtyards, 9.1% hope to collaborate with other households in courtyard animal husbandry, 4.6% prefer contracting with large family operations of animal husbandry, and 1.8% prefer contracting with animal husbandry enterprises.

3.2.2 *Migrant Workers*

As nonfarm employment and businesses gradually become the main economic activities of migrant families, over 50% of the migrants have been migrant workers. In the survey, 52.5% of the respondents said that they went out of hometown to work in 2015. According to the survey, in 2015, households without migrant workers account for 16.5% of the total, and those with one, two, and three and more migrant workers account for, respectively, 43.2%, 27.1%, and 13.1% of the total. A large proportion of migrant households have migrant workers. But the number of migrant workers is not high in comparison with the average family size of migrant households, i.e., 5.21 people. The income as migrant workers accounts for a large proportion of people's total income. In 2015, the percentage was less than 50% for 30.2% of the respondents, 50–99% for 33.9% of the respondents, and 100% for 35.8% of the respondents.

When asked about the stability of income derived from working out of hometown, 89.3% of the respondents chose "unstable," only 9% chose "stable," and 1.7% were not sure. Answers vary in different groups of people based on the nature of their jobs. The percentages of people choosing "stable" are, respectively, 33.3% and 22.2% in skilled labors and people in business. Only 9.2% of unskilled workers considered their income stable. Despite the fact that wages account for a large proportion of migrant's household income and contribute significantly to raising their household income, the instability of income from working out of hometown has had a negative impact on the efforts to boost the income of migrant families.

Of the respondents, 60.2% said that it was hard to find a job, 31.4% said it was not very difficult, and 8.4% said they had no difficulties finding a job at all. While seeking employment, 69.5% of the respondents have not been discriminated against because of being migrants, 14.1% said that they were occasionally discriminated against, and only 6.9% said that they were often discriminated against. The data shows that the migrants have well integrated into the local society and have not encounter discrimination while seeking employment.

Concerning channels of finding employment, 83.6% of the respondents relied on themselves, 15.2% turned to relatives for help, and only 0.4% were organized by government and labor agencies. The data shows that the migrants still rely on limited traditional channels to find nonfarm jobs, without making effective use of agencies in the labor market. Unlike in other parts of China, only a small number of migrants have been soliciting the help of relatives. As a result, the migrants have access to limited channels of employment and job information, causing low chances of getting the job. Government organizations have failed to take effective measures to help the migrants, which suggests the lack of long-term tracking of migrant's life and business after the resettlement. When asked about the desirable channels to find employment, 55.6% of the respondents preferred government organizations, 32.6% chose to rely on their own efforts, 6.6% preferred recommendation by relatives, and only 4.7% preferred market-based labor agencies. The people have shown great trust in and a strong demand for government organizations in finding employment. On the other hand, it shows that the migrants are not used to finding nonfarm jobs through market approaches.

In order to better help the migrants find nonfarm jobs, in advancing the ecological migration project, establishment of industrial parks and zones should be considered an important measure. Industrial parks and zones near relocation areas will enable these areas to develop by characteristic industries and enhance the impetus for sustainable development so as to lift the migrants out of poverty. According to the survey, 49.9% of the relocation areas are adjacent to industrial parks and zones, 29.9% of which are tourism zones, 10.7% are agricultural parks, and 10.1% are industrial parks. But only a small percentage of migrants, 22.9% of the migrants working out of home, earn wages from working in these parks and zones. It shows that the industrial parks and zones near relocation areas have only created limited jobs, which cannot satisfy the migrant's need for jobs. Of all the migrants who have been working out of home, 78.8% hope to find a job in nearby industrial parks and zones. Apparently, the existing industrial parks and zones can only create about one fourth of the jobs required.

As nonfarm jobs pose high requirement for migrant's kills, it is important for government to organize labor skill training to improve their skills while alleviating poverty so as to improve the jobs for migrants in terms of both quantity and quality. Major problems in this regard include inadequate training sessions and unsatisfactory results of training. Only 21.6% of the migrants have attended skill training organized by the government, and 41.4% of these people said that the training provided little or no assistance, 17.8% considered the training not so satisfactory, and 40.8% said that the training had been of some or great help.

3.3 Migrants' Demands for Policies on Production

A basic approach to solving problems emerging in the resettlement of government-organized migrants is to introduce and implement relevant policies. Constant efforts are required to increase financial input, improve facilities, boost industrial development, optimize the allocation of resources, and introduce new mechanisms. Therefore, the migrants have shown a strong demand for the following policies:

First, industrial policies to promote a diversified economy.

Industrial development is an important aspect of development-oriented poverty alleviation. While planning for resettlement of the ecological migrants, the government has designed policies for industrial development of relocation areas. Concerted efforts have been made to promote industrial development, build housing for migrants, improve infrastructure, and develop public facilities, giving priority to characteristic agriculture and employment services. The migrants believing that animal husbandry, plant production, employment services, services industry, processing industry, and tourism industry can boost their income, respectively, accounting for 63.3%, 60.9%, 21.9%, 9.6%, 9.3%, and 3.6% of the total. Overall, the migrants prefer boosting their income by animal husbandry and plant production.

According to the migrants, barriers to further agricultural development in relocation areas include lack of development funds (83.2%), shortage of farmland (54.0%), shortage of water (29.1%), poor quality of farmland (21.2%), shortage of labor (13.2%), inadequate irrigation facilities (15.6%), lack of economic knowledge (7.5%), and limited sales channel (5.6%). As a result, problems have persisted, such as small area of farmland per person, limit to the use of irrigation water, inadequate infrastructure, and farmers unable to command the techniques required in water-saving agriculture and protected agriculture. The biggest bottleneck that limits the sustainable development of migrants' agricultural operation is shortage of funds.

As for the further development of employment services and entrepreneurship in relocation areas, major barriers include lack of channels to find employment and start businesses (24.9%), lack of skills (23.9%), lack of access to information about relevant industries (18.0%), and inadequate ability to acquire new knowledge and skills (10.8%). Consequently, the sustainable development of employment services has been significantly undermined by slow progress in development and inadequate ability to create jobs of industrial parks, mismatch between migrants' quality and the requirements of the employment services industry, and limited channels to find employment and other problems.

Concerning the measures to be taken by the government to promote diversified industrial development and boost migrants' income, 74.5% of the respondents believed that enterprises processing characteristic agricultural produce and family-run businesses should be established in settlements to provide job opportunities for women and middle-aged and old men; 56.9% believed that the government should focus on local products with competitive edge and promote the development of agricultural products processing industry and services industry; 41.6% believed that the migrants should be encouraged to set up agricultural supplies stores, convenience stores, and other businesses catering to rural residents; 27.4% believed that efforts should be made to introduce or foster leading enterprises and engage them in industrial development of relocation areas by establishing a mechanism for industrial development and a business model involving enterprises, bases, and rural households.

The migrants calling for immediate assistance from the government in areas such as preferential credit policies, preferential industrial policies, employment and skill training, technical guidance, market information, and industrial guidance, respectively, account for 61.2%, 27.3%, 26.6%, 21.7%, 17.8%, and 13.0% of the total.

In short, the ecological migrants have been resettled, but their foundation for further development remains weak. Factors such as capital, land, and labor have not been well coordinated in economic development after the migration. As most of the migrants are organized by the government to migrate, they have great dependence on the local government in finding ways to raise their household income. The local governments, however, have only drafted relevant plans and resettled the migrants, but do not have the necessary funds, technology, and management tools to ensure sustainable economic development in relocation areas. As a result, in pursuit of economic development of relocation areas, the migrants tend to be dependent on local governments, while local governments tend to be dependent on financial input of the central government, leading to inadequate internal impetus for economic development after the migration and limited participation of the non-government sector.

Second, credit policies to address shortage of funds.

The biggest challenge for migrants in their lives and business operation is shortage of funds. The problem is attributable to many causes, such as expenditure during the relocation, investment in agricultural operation and entrepreneurship after the migration, unbalanced consumption structure, and so forth. According to the survey, since the migrants' friends and relatives are mostly also migrants facing similar financial problems, people tend to turn to loans to collect funds for agricultural operation. Therefore, the migrant households have a strong demand for credits.

Concerning credit policies to address shortage of funds, the migrants hope that the government can make it easier for them to get loans (91.1%), enhance financial support (62.5%), simplify the loan process (59.5%), encourage banks to take the initiative to offer loans (22.0%), expand village-level mutual funds (13.8%), and provide more channels to obtain loans (33.5%).

A major problem in granting credits to migrants is that the risk control policies of financial institutions on microfinance are not in line with poverty alleviation policies. Financial institutions, as creditors, impose strict requirements on risk control in relation to credit rating and granting, and they are reluctant to assume responsibility for credits to migrants. For the government, it is an important task to provide migrants with necessary funds so as to ensure the effect of migration. The migrants, however, tend to mistake poverty alleviation loans as relief funds. Therefore, all relevant parties must have a unified understanding of credits to migrants. Financial institutions should be more tolerant while controlling risks, the government should increase its effort to ensure precise registration of the poor and exercise strict supervision, and the migrants must change their mindset and have a clear understanding of their responsibility.

Third, agricultural policies on government subsidies.

Concerning agricultural policies, the migrants believe that subsidies are of much help to them in raising their household income. When asked about the policies that have played a significant role in boosting their household agricultural operation, the respondents referred to subsidies for protected agriculture, subsidies for training of a new generation of farmers, awards and subsidies for crop production, subsidies for agricultural businesses and agricultural service entities, subsidies for drip irrigation under membrane, and subsidies for safety technical inspection and insurance of agricultural machinery and equipment, respectively, accounting for 66.7%, 19.7%, 17.3%, 15.7%, 13.7%, and 11.0% of the total.

The rural households intending to raise cattle, sheep, poultry, and silkworms in the short term, respectively, account for 87%, 57%, 9.5%, and 0.7% of the total, and those intending to plant corns, goji berries, grapes, TCM herbs, and clover, respectively, account for 54.0%, 51.8%, 15.7%, 4.0%, and 1.8% of the total.

After the relocation, the migrants have better conditions for agricultural operation, but they are still bothered by outdated model of production, small area of farmland, low agricultural income despite considerable investment, shortage of agricultural facilities, deficiencies in agricultural techniques, and other problems. Some relocation areas have achieved development of characteristic agriculture, such as grape planting, but the overall development of characteristic agriculture in the areas covered in the survey has been slow. Further efforts are required to promote the good practice of demonstration and pilot areas of ecological migration in more areas.

Fourth, policies on government assistance in employment.

In view of the difficulties in finding jobs in nearby places and out of hometown, the migrants hope that the government can take multiple measures to facilitate the development of employment services. Of all the respondents, 49.4% believed that an incentive mechanism should be created to encourage enterprises in need of labor, such as large and medium-sized SOEs, collective enterprises, private enterprises, and other enterprises, to hire migrant workers directly from relocation areas; 46% believed that the government should enhance their cooperation with enterprises and other organizations in need of labor and establish a platform of employment information; 19.3% believed that efforts should be made to foster spearhead of the employment services industry so as to give full play to the role of high-caliber people; and 19.1% believed that more employment agents were needed. Only 0.1% of the respondents believed that the people needed to improve their knowledge and skills. The data shows that the migrants have great dependence on government and high-caliber people, without paying much attention to improving themselves. The skills training sessions organized by the government are considered not relevant (59.9%), training only elementary skills (40.7%), too short (38.9%), and offering limited lessons (50.3%). Such training sessions, which fail to consider the requirements of the market and the characteristics of the potential trainees, will not have much effect.

The migrants have a demand for policies on recruiting teachers and establishing schools. Concerning the education-related tasks that require immediate attention, 66.7% of the respondents referred to recruiting quality teachers, 61.7% referred to school building, and 46.9% referred to increased spending on education. The most needed skills, according to the respondents, include construction (56.9%), driving (41.6%), welding (36.1%), tailoring (30.6%), cooking (19.4%), and household services (16.7%).

Of all the respondents, 57.6% have intentions to start a business. In this regard, the migrants are mainly hindered by problems such as lack of funds (94.4%), lack of entrepreneurship projects (42.6%), lack of entrepreneur guidance (32.3%), and lack of entrepreneurship policies (27.7%).

Fifth, poverty alleviation policies: over 80% of the migrants believed that they should have access to precision-targeting poverty alleviation measures.

As mentioned, the migrant households are, based on migrant's self-assessment, classified into extremely poor households, households receiving subsistence allowance, households enjoying the "five guarantees," average households, and rich households. 84.2% of the respondents believed that they should have access to precision-targeting poverty alleviation measures. This opinion is shared by 97.5% of extremely poor households, 90.1% of households receiving subsistence allowance, 90% of households enjoying the "five guarantees," 76% of average households, and 12.5% of rich households. The data reveals a strong tendency toward absolute equality, equitable distribution of benefits and generalization of migration policies among the migrants. This has led to problems in implementation of poverty alleviation policies in the migration project. Among the households covered in the survey, 29.8% enjoy subsistence allowance, 9.7% have benefited from poverty alleviation by education, 2.8% have received vocational training, and 0.9% are entitled to supportive policies for characteristic industries. Among the rich households, 12.5% have benefited from supportive policies for characteristic industries and education policies meant to alleviate poverty. Among the average households, 11.3% have benefited from education policies meant to alleviate poverty, 8.2% enjoy subsistence allowance, 3.6% have benefited from policies on vocational training for migrant workers, and 1.3% have benefited from supportive policies for characteristic industries. The extremely poor households and households receiving subsistence allowance and the "five guarantees," however, have not had access to other poverty alleviation measures except for subsistence allowance and training for migrant workers. It is evident that the precision-targeting poverty alleviation policies have not been fully implemented in the ecological migration project.

The most in-demand profession is agricultural technician, accounting for 69.5% of the total. Financial and educational professionals rank second and third, accounting for 14.1% and 8.7% of the total, respectively. The demand for e-commerce, sales, and medical professionals accounts for less than 5% of the total, respectively. Although over 50% of the migrant households are no longer engaged in agricultural operation, the households in need of agricultural technicians still constitute nearly 70% of the total. This is an indication that during the relocation, the industrial policies have failed to consider the condition and demand of migrants.

The migrants also show a pressing need for agricultural projects to alleviate poverty. The demand for animal husbandry projects, plant production projects, education projects, skills training projects, water conservancy projects, deep processing projects, road projects, tourism projects, and e-commerce projects accounts for 35.3%, 34.0%, 10.1%, 9.0%, 4.0%, 2.7% 1.7%, 1.1%, and 0.7% of the total, respectively.

In view of the government's increasing efforts to alleviate poverty, the migrants have shown great confidence that they will cast off poverty and earn higher income. In this regard, 34.3% of the respondents were very confident, 45.5% were somewhat confident, 11.3% were slightly confident, and 6.1% were not confident.

Sixth, policies on ecological compensation: over 80% of the migrants entitled to ecological compensation require higher compensation.

According to the survey, 7.5% of the migrant households are entitled to ecological compensation to non-commercial forest, of which 54.3% believed that the compensation enabled a significant increase in their household income, 18.6% did not experience striking changes, and 27.2% believed that the compensation did not have any effect or that the effect was not significant. In order to enable the non-commercial forest to play a better role in boosting people's income, 85.2% of the respondents suggested raising the compensation, and 7.4% suggested promoting the development of under-forest economy.

The survey also shows that 3.1% of the respondents are entitled to grassland ecological protection subsidies and rewards. Of these households, 19.2% said that the rewards significantly increased their household income, 38.5% said that there was no significant change, and 42.3% said that the rewards did not have any effect or that the effect was not significant. In order to enable the grassland ecological protection subsidies and rewards to play a better role in boosting people's income, 84.6% of the respondents suggested increasing the subsidies, and 11.5% suggested promoting the development of grassland tourism. The survey shows that the ecological compensation has to some extent raised the income of migrant households, but the sustainability of the policy is undermined by migrant's excessive dependence on compensation.

As is evident from the above demands for policies, the migrant households have excessive dependence on supportive policies and subsidies. Meanwhile, there has been some deviation in implementation of some precision-targeting poverty alleviation policies. The limitations of government in making plans for industrial development and job market by administrative means have also emerged. The government is required to resettle the ecological migrants and help them overcome subsequent difficulties in production. But as the objects and targets of relevant policies are not properly set, and the government has failed to consider the role of the market and the non-government sector in formulation of policies, the policies are not as effective as thought. In advancing the migration project, the government must value the role of the market and the non-government sector and encourage the development of non-governmental organizations so as to address problems in migrant's economic activities.

Chapter 4
Analysis of "Low Settlement Rate" in the Ecological Migrant Settlements in Ningxia

Jianrong Fan

With years of practices, ecological migration has proven to be an important approach for Ningxia Hui Autonomous Region to accelerating the development of its poverty-stricken areas and is drawing greater attention. For the regions that the migrants were relocated from, the migration has brought them lots of benefits, with the ecological environment restored, the tension between land resource and population eased, the poverty rate reduced, and more potentials for development created. For the host places, the inflow of the ecological migrants has also enabled the effective utilization of the land resources and relieved workforce shortage in those areas. More importantly, it has also sped up coordinated development across Ningxia.

Nevertheless, over time, with the deepening of migration process, the implementation of ecological migration in Ningxia is also faced with increasing challenges. And some of them must be addressed immediately, such as self-initiated migration, industrial development, and the settlement rate among the settlers. One of the most acute problems among all these is the low settlement rate in the ecological migrant settlements.

"Low settlement rate" in an ecological migrant settlement refers to the low proportion of the migrants having settled in the ecological migrant settlement relative to the number of migrants expected to settle there under the related migration policy.

To understand the issue of low settlement rate of ecological migrants comprehensively, the research group conducted four surveys around the topic by random sampling successively in the four major migrant settlements located in Xijing, Yuanzhou, and Hongsipu districts in Ningxia from February to April 2016. Based on the primary sources collected thereby, the team intends to find out the root causes of the low settlement rate, yielding insights to the issue while proposing effective solutions in favor of ecological migration and precision-targeting poverty alleviation, so as to facilitate the migration process and the rapid and sound development

J. Fan (✉)
School of Economics, North Minzu University, Yinchuan, China
e-mail: nxlob@163.com

X. Wang (ed.), *Ecological Migration and Targeted Poverty Alleviation in Ningxia*, International Research on Poverty Reduction,
https://doi.org/10.1007/978-981-19-7888-3_4

of migrant settlements. We distributed 207 questionnaires in the four surveys with 100% of them returned and effectively answered.

4.1 Migrant Respondents' Perceptions of the Current Condition of Relocation

Since the implementation of ecological migration in Ningxia in 1980s, a variety of models of ecological migration have been developed, incubating a collection of diversified residential communities of ecological migrants scattered across the autonomous region. These include both the unitary communities that are populated exclusively by either ecological migrants or labor migrants, as well as the hybrid communities which accommodate both the two types of migrants (see Table 4.1).

We found through long-term study on ecological migration that self-initiated migration is one of the main reasons for the low rate of settlement in the communities of ecological migrants. Therefore, in this study, we only selected the policy-based communities as survey samples, which include two ecological migrant settlements relocated inside the county—Xingde Village in Qiqiang Town (an ecological migrant settlement in Xiji County, Ningxia) and Yuande Migrant New Village in Touying Town (an ecological migrant settlement in Yuanzhou District, Ningxia); and one labor migrant settlement relocated inside the county—Fengze Migrant New Village in Yuanzhou District, Ningxia; and one hybrid ecological migrant settlement relocated outside the county—Lujiayao Migrant Village in Hongsipu District, Ningxia.

By theoretical review and detailed analysis, we have gained some in-depth knowledge on the low settlement rate in ecological migrant settlements and put the factors that may affect settlement rate into five categories: first, the residents who should have settled down in the new community as per the policy but did not; second, the

Table 4.1 Major types of ecological migrants in Ningxia

		Migration within the same county	Migration outside the county
Policy-based migrants	Ecological migrants	Policy-based ecological migrants relocated inside the county	Policy-based ecological migrants relocated outside the county
	Labor migrants	Policy-based labor migrants relocated inside the county	Policy-based labor migrants relocated outside the county
Self-initiated migrants	Ecological migrants	Self-initiated ecological migrants relocated inside the county	Self-initiated ecological migrants relocated outside the county
	Labor migrants	Self-initiated labor migrants relocated inside the county	Self-initiated labor migrants relocated outside the county

residents who had settled in a community in compliance with the policy but soon relocated themselves once again; third, the residents who do not stay in the community but work elsewhere; fourth, the potential migrants. Namely, the respondents who have the willingness to relocate again after the resettlement but yet to do it because of various restrictions; and fifth, the house lessors or sellers, which refers to the residents who enjoyed the preferential policies as ecological migrants but do not want to live in the community thus try to profit from what they have obtained.

4.1.1 On "Those Who Should Have Resettled in but didn't"

4.1.1.1 Analysis on the Residents Who Should Have but Did not Relocate and the Reasons

(1) Judgment on resettlement participation

As can be inferred from Table 4.2, more than 50% of the respondents believe that those who should be subject to relocation have all been resettled. But there are also 10–40% of the respondents think that there are some residents who should have relocated but did not. This indicates that people who did not participate in relocation are important contributors to the low settlement rate.

(2) Analysis on why some residents did not relocate

It can be read from Table 4.3 that the respondents from different residential communities have different observations in terms of why some villagers did not leave their habitual homes even though they are asked to. In the ecological migrant settlement in Xiji County, the option "dissatisfaction" ranks the first, since it is a community that accommodates ecological migrants within the county. They were relocated from the areas where the ecological environment is relatively fragile and were resettled in the neighboring regions along railways, driveways, or water resources, thus what really changed for them is the housing condition, while the circumstance of development did not change much. Consequently, some of the residents felt that the relocation is "not satisfactory" and refused relocation. As high as 80% of the respondents in one community surveyed believed that this was the crucial reason. The reason ranking the second is "family poverty." Even though the ecological migration

Table 4.2 Respondents' judgment on the residents who should or should not be relocated (%)

	Should	Should not	Unsure
The ecological migrant settlement in Xiji County	67.3	9.6	23.1
The labor migrant settlement in Yuanzhou District	50.0	38.0	12.0
The ecological migrant settlement in Yuanzhou District	87.5	12.5	0.0
The hybrid ecological migrant settlement in Hongsipu District	77.6	14.3	8.2

Table 4.3 Respondents' observations on why the non-participants refuse to relocate (%)

	Family poverty	Not eligible	Self-initiated relocation	Unwillingness	Dissatisfaction
The ecological migrant settlement in Xiji County	60.0	20.0	20.0	40.0	80.0
The labor migrant settlement in Yuanzhou District	5.3	21.1	0.0	5.3	5.3
The ecological migrant settlement in Yuanzhou District	14.3	14.3	0.0	14.3	0.0
The hybrid ecological migrant settlement in Hongsipu District	14.3	28.6	0.0	14.3	14.3

Note This is a multiple-choice question

planning stipulates that each village should be relocated as an integrity, there were still some impoverished residents unable to afford the relocation fee of 12,800 yuan for the house and related expenses, thus cannot be relocated. This phenomenon also caused low rate of settlement. While in both the hybrid ecological migrant settlements in Hongsipu District and the labor migrant settlements in Yuanzhou District, people choosing "noncompliance with the policy" outnumbers the rest. Because of the loose household inspection procedure, some over-populated migrant families managed to bring more members than the limited living space could accommodate. The poor living condition and uncertain prospects drove these people to return.

4.1.1.2 Impact of These Should-Have Settlers on the Development of Community

(1) On whether these should-have settlers have an impact on the settlement

As can be seen from Table 4.4, the respondents have different opinions about whether the settlers not having moved in had an impact on the development of community, but their answers have certain regularity. In the ecological migrant settlements designed for land resettlement, it is generally believed that the impact is little,

Table 4.4 Respondents' observations on whether the non-participants have an impact on the settlement (%)

	Yes	No
The ecological migrant settlement in Xiji County	0.00	100.00
The labor migrant settlement in Yuanzhou District	52.60	47.40
The ecological migrant settlement in Yuanzhou District	14.30	85.70
The hybrid ecological migrant settlement in Hongsipu District	14.30	85.70

with the proportion all exceeding 85%. But a different opinion was given by the community designed for labor migration, as the percentage of the respondents who think that there is an impact and is not both stand at 50%.

(2) Analysis on the reason for the impact

It can be inferred from Table 4.5 that the major areas affected in the eyes of the respondents vary from community to community. In the ecological migrant settlement in Yuanzhou District, the respondents believe that the residents who did not settle in will affect "agricultural production," "public security of community," and "children's schooling"; in the labor migration community in Yuanzhou District, the major concerns concentrate at "public security of community" and "community management"; and the hybrid ecological migrant settlement in Hongsipu District worries about "public security of community" and "children's schooling" most. Due to the existence of the residents who did not settle in against the policy, the communities suffered from an inflow of excessive external population, bringing acute challenges to public security. Against the ordinary procedure, the children of these should-have settlers also did not register at the community schools for enrollment, causing low enrollment rate and attendance rate in these schools thus affecting their daily operation. Consequently, it created a vicious circle and the education resources were wasted. Therefore, the subject drew the attention of the respondents.

(3) Analysis on why there is no impact on the development of communities

Table 4.6 shows that the respondents chose "no impact" for the same reason although they live in different communities, namely the residents of each community have "no impact on each other." It not only reveals once again the characteristic of the existing ecological migrant settlements, but also explains why some residents believe that no impact will be exerted by the non-participants. Ever since the implementation of the household contract responsibility system in China, the agricultural production in rural areas has been carried out by every household independently, making these rural households less and less bonded. It is especially true in the mountainous areas of southern Ningxia, where the agricultural production is deeply weather-dependent.

Table 4.5 Impacts of the non-participants on the settlement (%)

	Agricultural production	Public security of community	Children's schooling	Community management
The ecological migrant settlement in Xiji County	0.0	0.0	0.0	0.0
The labor migrant settlement in Yuanzhou District	0.0	80.0	0.0	40.0
The ecological migrant settlement in Yuanzhou District	100.0	100.0	100.0	0.0
The hybrid ecological migrant settlement in Hongsipu District	0.0	100.0	100.0	0.0

Note It is a multiple-choice question

Table 4.6 Analysis on why non-participants have no impact on the development of communities (%)

	No impact on each other	No organized activities or events in the community	No agricultural production	No institution in the community
The ecological migrant settlement in Xiji County	80.0	0.0	0.0	40.0
The labor migrant settlement in Yuanzhou District	42.1	0.0	15.8	5.3
The ecological migrant settlement in Yuanzhou District	85.7	0.0	0.0	0.0
The hybrid ecological migrant settlement in Hongsipu District	71.4	14.3	14.3	0.0

Note This is a multiple-choice question

Table 4.7 The respondents' observations on whether some migrants relocated again (%)

	Yes	No	Unsure
The ecological migrant settlement in Xiji County	100.0	0.0	0.0
The labor migrant settlement in Yuanzhou District	20.0	70.0	10.0
The ecological migrant settlement in Yuanzhou District	17.9	82.1	0.0
The hybrid ecological migrant settlement in Hongsipu District	24.5	61.2	14.3

4.1.2 On "The Migrants Who Settled in and Relocated Again"

4.1.2.1 Judgment on the Existence of Migrants Who Settled in and Relocated Again

As can be told from Table 4.7, all the communities surveyed have residents believing that there are migrants who moved again after resettlement, with the statistic in the ecological migrant settlement in Xiji County stands out most, which hits a percentage of 100%. The data in the rest communities, though relatively lower, can also testify the existence of the outflow of the resettled migrants—another contributor to the low settlement rate in ecological immigrant settlements. As what we found during the field trips, some of the relocated households have deserted their houses, leaving yards overgrown with weeds, gates closed tight, and even doors and windows blocked by bricks. There are also "migratory birds"-like migrants, who only showed up in the community when they were required, and then returned to the location of the registered household or other places.

4.1.2.2 Observations on the Migrants Who Moved Again

Table 4.8 tells us that despite the different development status of these communities, their residents had similar opinions about what kind of people would choose to relocate again. The replies could be generally summarized as those who "see things differently from others" and who "have jobs elsewhere all year long."

4.1.2.3 Analysis on Why Some Migrants Relocated Again

Table 4.9 shows that the reasons of re-relocation vary from community to community, but there are some common themes. Many respondents cited "unfavorable economic conditions" as the reason. This is because as more of these projects are approved and implemented, locations suitable for such purpose have become harder to find, and more sites are chosen despite lacking what is needed to support the migrant population. It is a common feeling shared by the ecological migrants. For the ecological

Table 4.8 Respondents' observations on the migrants who relocated again (%)

	See things differently from others	Have jobs elsewhere all year long
The ecological migrant settlement in Xiji County	38.5	59.6
The labor migrant settlement in Yuanzhou District	20.0	20.0
The ecological migrant settlement in Yuanzhou District	40.0	60.0
The hybrid ecological migrant settlement in Hongsipu District	16.7	8.3

Note This is a multiple-choice question

migrant settlements mainly built for land resettlement, the answers mainly focused on "demands can't be met" and "seek for better opportunities." While for the ecological migrant settlement in Xiji County, a new community that has just accomplished resettlement, "have better places to go" is another important reason, which is applicable to the circumstances of self-initiated migration and education migration for children's schooling.

Table 4.9 Respondents' analysis on why the migrants relocated again (%)

	Poor condition of development	Have better places to go	Demands can't be met	Seek for better opportunities
The ecological migrant settlement in Xiji County	42.3	61.5	38.5	57.7
The labor migrant settlement in Yuanzhou District	50.0	0.0	0.0	0.0
The ecological migrant settlement in Yuanzhou District	70.0	0.0	20.0	50.0
The hybrid ecological migrant settlement in Hongsipu District	25.0	0.0	33.3	25.0

Note This is a multiple-choice question

4.1.3 *On "The Residents Who Left to Work Elsewhere"*

4.1.3.1 Observations on the Residents Who Left to Be Migrant Workers and Why They Did so

(1) On the outflow of migrant labor force

Table 4.10 shows in all of the communities surveyed, over 79% of respondents believed that many in the community have left the settlement to find employment elsewhere. For rural families, the remuneration of migrant workers is one of the main sources of income. Given the poor conditions for agricultural production in the settlements, many have had to resort to working as wage-earners elsewhere.

(2) Reason analysis

As Table 4.11 reads, although the proportion differs slightly, "demands hardly satisfied" and "poor conditions for development" account for the first two reasons for the new occupants in the labor migrant settlement and ecological migrant settlement in Yuanzhou District and the hybrid ecological migrant settlement in Hongsipu District—where the relocation was conducted relatively early—to move away and seek employment in other regions. On the one hand, although migration ameliorated people's living conditions, the general environment remained the same. On the other hand, as the relocated households were given lesser or even no farmland, the new residents found nowhere to utilize their traditional farming methods thus felt unfulfilled. As for the relatively new ecological migrant settlement in Xiji County, although the social conditions for development have been improved relatively, the improvement is not thorough enough. Therefore, the work-oriented out-migration here also has two main features. First, the migrant workers who accomplished the "relocation" following the instruction of policy found that they would benefit more from working outside than staying at home, thus continued to work elsewhere. Second, some residents found that the relocation did not meet their expectation and working outside was more profitable; thus, they chose to leave the community and work elsewhere.

Table 4.10 Respondents' observations on the outflow of migrant labor force in the community (%)

	Many	Not many	Unsure
The ecological migrant settlement in Xiji County	94.2	0.0	5.8
The labor migrant settlement in Yuanzhou District	88.0	0.0	12.0
The ecological migrant settlement in Yuanzhou District	91.1	8.9	0.0
The hybrid ecological migrant settlement in Hongsipu District	79.6	16.3	4.1

Table 4.11 Reasons cited for why some migrants left the settlement to work elsewhere (%)

	Unfavorable economic conditions	Already had stable employment elsewhere	Demands hardly satisfied	To seek better opportunities
The ecological migrant settlement in Xiji County	15.4	53.8	23.1	7.7
The labor migrant settlement in Yuanzhou District	18.0	0.0	46.0	0.0
The ecological migrant settlement in Yuanzhou District	37.5	0.0	62.5	14.3
The hybrid ecological migrant settlement in Hongsipu District	20.4	8.2	61.2	2.0

Note This is a multiple-choice question

4.1.3.2 Impacts of the Migrant Workers on the Ecological Migrant Settlement

(1) Judgment on whether there is impact or not

Table 4.12 tells us that in all the communities surveyed, the residents who thought there was no impact outnumbered those who thought there was, and the margin was largest in the labor migrant settlements.

(2) Analysis on why there is impact

From Table 4.13, we can see that the outflow of migrant labor force could affect ecological migration in several ways, to varying degrees in different communities. Their influence on the ecological migration community in Yuanzhou District is comprehensive, with "children's schooling" ranking first and "agricultural production" and "public security of the society" both coming at the second. In the hybrid ecological migrant settlement in Hongsipu District, the option "children's schooling" is most frequently cited. In the ecological migrant settlement in Xiji County,

Table 4.12 Respondents' judgment on the impacts of the migrant workers on the settlement (%)

	Yes	No
The ecological migrant settlement in Xiji County	38.5	61.5
The labor migrant settlement in Yuanzhou District	4.0	96.0
The ecological migrant settlement in Yuanzhou District	32.1	67.9
The hybrid ecological migrant settlement in Hongsipu District	44.9	55.1

Table 4.13 Analysis on why the outflow of migrant labor force has impact on the development in a settlement (%)

	Agricultural production	Public security of the society	Children's schooling	Regular management	Development of community
The ecological migrant settlement in Xiji County	40.0	0.0	80.0	0.0	0.0
The labor migrant settlement in Yuanzhou District	0.0	0.0	0.0	0.0	0.0
The ecological migrant settlement in Yuanzhou District	66.7	66.7	72.2	50.0	22.2
The hybrid ecological migrant settlement in Hongsipu District	9.1	0.0	68.2	9.1	4.5

Note This is a multiple-choice question

"children's schooling" ranks the first and "agricultural production" ranks the second. Evidently, the education of children is one of the greatest concerns of the ecological migrants. It is common for the migrant workers to take their school-aged kids traveling with them, leaving the number of students to be received by the local schools shrinking gradually, a harbinger of the decline of these already existed schools.

(3) Analysis on why there is no impact

Table 4.14 explains why some residents think that the exodus of migrant labor force has no impact on the development of community, mainly focusing on two aspects: firstly, "neither affects the other," and secondly, "settlement has no agricultural production."

Table 4.14 Analysis on why the outflow of migrant labor force has no impact on the development of community (%)

	Neither affects the other	Settlement has no agricultural production
The ecological migrant settlement in Xiji County	62.5	0.0
The labor migrant settlement in Yuanzhou District	79.2	31.3
The ecological migrant settlement in Yuanzhou District	68.4	13.2
The hybrid ecological migrant settlement in Hongsipu District	82.4	23.5

Note This is a multiple-choice question

4.1.4 On "The Potential Migrants"

4.1.4.1 Knowledge of the Respondents on Models of Migration and the Model They Adopted

(1) Knowledge acquired on models of migration

Table 4.15 shows that knowledge acquired by the respondents on models of ecological migration is unsatisfactory enough. For the labor migrant settlement, nearly 90% of the respondents have certain knowledge. And for the ecological migrant settlement in Xiji County, the proportion of the residents "know something about them" is higher than those who "know nothing about them." In the ecological migrant settlement in Yuanzhou District, a community relocated relatively earlier, people "know about them" accounts for apparently higher percentage than that in the community of Xiji County, whose relocation was completed relatively more present. Surprisingly, over 90% of the respondents in the hybrid ecological migrant settlement in Hongsipu District admitted that they did not know about migration models.

(2) The respondents' knowledge on their own migration model

Table 4.15 Respondents' knowledge on the models of ecological migration (%)

	Know something about them	Know nothing about them
The ecological migrant settlement in Xiji County	51.9	48.1
The labor migrant settlement in Yuanzhou District	88.0	12.0
The ecological migrant settlement in Yuanzhou District	64.3	35.7
The hybrid ecological migrant settlement in Hongsipu District	6.1	93.9

Table 4.16 Respondents' knowledge on the model of their own migration (%)

	Know about it	Do not know about it
The ecological migrant settlement in Xiji County	90.4	9.6
The labor migrant settlement in Yuanzhou District	90.0	10.0
The ecological migrant settlement in Yuanzhou District	83.9	16.1
The hybrid ecological migrant settlement in Hongsipu District	36.7	63.6

Table 4.17 Respondents' knowledge about the exact model of their own migration (%)

	Ecological migration	Labor migration
The ecological migrant settlement in Xiji County	100.0	0
The labor migrant settlement in Yuanzhou District	13.3	86.7
The ecological migrant settlement in Yuanzhou District	100.0	0
The hybrid ecological migrant settlement in Hongsipu District	100.0	0

It can be told from Table 4.16, expect for the residents surveyed in the hybrid ecological migrant settlement in Hongsipu District, the percentage of the respondents of the rest communities who "know about" their own migration model all reached 80%. And as Table 4.17 shows, those who "knew about" the models were clearly aware of the model of their own relocation. But as the settlement in Hongsipu District is a hybrid one, only one third of the residents surveyed reported "knowing about" the migration model applicable to their own move.

4.1.4.2 The Potential Willingness to Migrate, and Reasons, Among the Respondents

(1) Potential willingness to migrate

As can be seen from Table 4.18, the proportion of "unwilling" in the communities surveyed all exceeds 50%. Relatively, the number of people who are willing to relocate once again in the labor migrant settlement is far lower than that of the ecological migration communities of other types, accounting for only 80%. In the ecological migration communities, the longer the residents have resettled, the fewer people have willingness to migrate again. The hybrid ecological migration community in Hongsipu District is the earliest (10.2%); the ecological migrant settlement in Yuanzhou District (12.5%) is the second earliest; and the third earliest, namely the latest one is Xiji County ecological immigrant settlement, whose "willingness" to relocate again registers as high as 40.4%.

(2) Reasons of the potential willingness for migration

Table 4.18 Respondents' judgments on whether they are willing to migrate again (%)

	Willing	Unwilling	Unsure
The ecological migrant settlement in Xiji County	40.4	51.9	7.7
The labor migrant settlement in Yuanzhou District	8.0	70.0	22.0
The ecological migrant settlement in Yuanzhou District	12.5	82.1	5.4
The hybrid ecological migrant settlement in Hongsipu District	10.2	71.4	18.4

Table 4.19 Respondents' reasons for being willing to migrate again (%)

	Poor conditions in current settlement	Good government policy	Seeking new opportunities	Government relocation
The ecological migrant settlement in Xiji County	38.1	0.0	61.9	0.0
The labor migrant settlement in Yuanzhou District	100.0	0.0	100.0	0.0
The ecological migrant settlement in Yuanzhou District	42.9	0.0	100.0	0.0
The hybrid ecological migrant settlement in Hongsipu District	0.0	20.0	40.0	20.0

Note This is a multiple-choice question

As can be seen from Table 4.19, the ecological migrant and labor migrant settlements relocated within the county have shared reasons for re-migration. The reason ranking first is "to seek new opportunities," and the second is "poor living conditions." The hybrid ecological migrant settlement relocated outside of the county reflects more diversified choices, with "to seek new opportunities" ranking the first, and "good government policy" and "government relocation" ranking the second.

(3) Reasons of willingness for migration

It can be read from Table 4.20 that although different communities have different choices, the reasons focus on two aspects: "too troublesome" and "too risky".

4.1.4.3 Preferred Model of Second Migration

(1) Choice of models

By Table 4.21, we can tell that different communities have different choices on relocation model for the next possible migration. The ecological migrant settlement

Table 4.20 Respondents' reasons for being unwilling to migrate again (%)

	Too risky	Other options	Unfavorable policy	Poor location	Too troublesome
The ecological migrant settlement in Xiji County	33.3	0.0	0.0	0.0	81.5
The labor migrant settlement in Yuanzhou District	25.7	8.6	0.0	0.0	40.0
The ecological migrant settlement in Yuanzhou District	32.6	6.5	0.0	0.0	23.9
The hybrid ecological migrant settlement in Hongsipu District	57.1	0.0	25.7	34.3	82.9

Note This is a multiple-choice question

in Xiji County and the labor migrant settlement in Yuanzhou District have clearer and more consistent choices, with the majority choosing "ecological migration" over "labor migration." The choices cited by the communities in Yuanzhou District and Hongsipu District are relatively unclear, with the proportion of residents who are "not sure" (others) about what model to choose higher than those who specified the exact models they preferred.

(2) Reason analysis on why ecological migration is preferred over labor migration

Table 4.21 Preferred model of second migration (%)

	Ecological migration	Labor migration	Others
The ecological migrant settlement in Xiji County	61.9	38.1	0.0
The labor migrant settlement in Yuanzhou District	75.0	25.0	0.0
The ecological migrant settlement in Yuanzhou District	28.6	28.6	42.9
The hybrid ecological migrant settlement in Hongsipu District	20.0	20.0	60.0

(a) Reason for choosing ecological migration

Table 4.22 indicates that the reasons for the ecological migrants in Yuanzhou District to choose "ecological migration" again are diversified, as almost all the options have been ticked by the respondents except "better policy." For the ecological migrants in Xiji County, although the reasons are also diverse, their priorities are much clearer. What ranking at the top is "have land for resettlement," an option to meet the basic demand of farmers "to have their own land." At the second place is "secured housing," as "there will be no worries when you have your house and land." The third is "greater adaptability," since they will still be engaged in farming. For the labor migrants in Yuanzhou District, they also yearn for "ecological migration" because for farmers, having their own lands is their best guarantee. For the hybrid ecological migrants in Hongsipu District, they chose "housing security" and "better policy" since they were already benefited from ecological migration; thus, they hoped to get more from the next migration.

(b) Reasons for not choosing labor migration

Table 4.23 tells us that for the ecological migrants in Xiji County, the reason that they will not choose "labor migration" lies in "no land, more concerns," a reflection of the "subconsciousness" of farmers. For the labor migrants in Yuanzhou District, the reason is "higher risk of unemployment," reflecting the fact that some migrants

Table 4.22 Reasons for choosing ecological migration first for a second migration (%)

	Have land for resettlement	Can find employment	Higher adaptability	Better policy	Secured housing
The ecological migrant settlement in Xiji County	100.0	0.0	30.8	0.0	38.5
The labor migrant settlement in Yuanzhou District	100.0	0.0	0.0	0.0	0.0
The ecological migrant settlement in Yuanzhou District	100.0	100.0	100.0	0.0	100.0
The hybrid ecological migrant settlement in Hongsipu District	0.0	0.0	0.0	100.0	100.0

Note This is a multiple-choice question

Table 4.23 Reasons for not choosing labor migration first for a second migration (%)

	Higher risk of unemployment	Uncertain identity	No land, more concerns
The ecological migrant settlement in Xiji County	0.0	0.0	100.0
The labor migrant settlement in Yuanzhou District	100.0	0.0	0.0
The ecological migrant settlement in Yuanzhou District	100.0	100.0	0.0
The hybrid ecological migrant settlement in Hongsipu District	0.0	0.0	0.0

Note This is a multiple-choice question

could not adopt to the migration model they followed. Also, it reflects that some migrants worried that their background of education and skills might not enable them to be adapted to the new environment and find employment, thus rejecting a second time "labor migration." For the migrants in Yuanzhou District, they felt that they would suffer from "higher risk of unemployment" and "uncertain identity," also a reflection on the existing problems in labor migrant settlement.

(3) Reason analysis on why labor migration is preferred over ecological migration

(a) Reasons for choosing labor migration

As can be seen in Table 4.24, except the labor migrant settlement in Yuanzhou District, the other settlements in Xiji County and Hongsipu District all cite "more employment opportunities" as a reason for them to consider labor migration, which shows that migrants do have good expectations on this model. For the ecological migrant settlement in Yuanzhou District, their choices were more diversified. "Greater support from the government" is also put into consideration, together with "more employment opportunities," demonstrating people's certain awareness on labor migration policies.

(b) Reasons for rejecting ecological migration

As can be seen in Table 4.25, the reasons why ecological migration is not their first choice for a second move differ from community to community. Those in the ecological migrant settlement in Xiji County worried about "limited opportunities for personal development," while the hybrid ecological migrant settlement feared of "limited income from agricultural production." For the community in Yuanzhou District, "limited income from agricultural production" and "difficulty implementing policy" are the two main concerns of the respondents.

Table 4.24 Reasons for preferring labor migration for the second migration (%)

	Secure housing	More employment opportunities	Greater support from the government
The ecological migrant settlement in Xiji County	0.0	100.0	0.0
The labor migrant settlement in Yuanzhou District	0.0	0.0	0.0
The ecological migrant settlement in Yuanzhou District	50.0	100.0	100.0
The hybrid ecological migrant settlement in Hongsipu District	0.0	100.0	0.0

Note This is a multiple-choice question

Table 4.25 Reasons why ecological migration is not top choice for second migration (%)

	Limited development opportunities	Limited income from agricultural production	Policy implementation difficulties
The ecological migrant settlement in Xiji County	100.0	0.0	0.0
The labor migrant settlement in Yuanzhou District	0.0	0.0	0.0
The ecological migrant settlement in Yuanzhou District	0.0	100.0	50.0
The hybrid ecological migrant settlement in Hongsipu District	0.0	100.0	0.0

Notes This is a multiple-choice question

4.2 Analysis of the Causes of Low Settlement Rate in Ecological Migrant Settlements

Ecological migration is a typical government action. Ningxia ecological migration project has been effective in alleviating poverty and mitigating the ecological environment degradation in the southern region, and the Government is also exerting utmost efforts to ensure ecological migrants' ability to move, as well as their stability and prosperity. However, due to multiple factors, such as the low quality of ecological migrants, inadequate management measures, strong speculative mentality, and poor community development environment, and especially since some people start to consider future development after solving the problem of food and clothing and having their stomach fed, they begin to complain about the lack of a stable economic source. Feelings of incompatibility with the production methods of their income sources start to spread among ecological migrant settlements, contributing to the growing problem of low settlement rates.

4.2.1 Increasingly Prominent Location Selection Issues, Poor Subsequent Development Environment

In order to ensure that ecological migrants can live and work in peace and contentment, the Government has provided favorable housing conditions and a certain amount of land resources for them, preferable infrastructure for community development, and some support for the follow-up industrial development for the development of ecological migrants.

As the ecological migration work progresses, its location selection issue is growing increasingly conspicuous, and locations with a decent development environment, that is near water, along the road, and next to the city, are becoming ever more difficult to find, leaving some ecological migrant settlements without a robust follow-up industry. It is not that the local governments focus only on infrastructure development and not on the development and nurturing of follow-up industries, but that it is too challenging. In particular, as the macro economy enters a new normal, the previously satisfactory development environment of ecological migrant settlements also deteriorates. Land is needed, but it is few in number, inferior in quality, and already under trusteeship; jobs are needed, but as economic development weakens, it is hard to find one.

As a result of poor community development (poor living conditions) (see Table 4.26), some of the ecological migrants have gone out in search of new development opportunities. In several of the communities surveyed, migrant workers were present in almost every household, with a high number of the elderly, women, and children remaining at home. The basic situation of the respondents is enough to demonstrate this phenomenon.

4.2.2 Low Level of Community Management, Low Efficiency of Supporting Facilities

Ecological migrant settlements are communities with high standards and a high starting point that the State has spent abundant human, material, and financial resources to construct, with tidy housing and decent infrastructure that the people

Table 4.26 Proportions of respondents who think the settlement offers few economic opportunities, by type (%)

	Xiji County	Yuanzhou District	Yuanzhou District	Hongsipu District
	Ecological migrants	Labor migrant	Ecological migrants	Mixed ecological migrants
Migrants that should have settled but have not yet	80.0	5.3	0.0	14.3
Migrants that have settled but do not live there	42.3	50.0	70.0	25.0
Migrant workers	15.4	18.0	37.5	20.4
Potential migrants	38.0	100	42.9	0

can see and appreciate. However, there are management issues unfavorable to the community management and development.

4.2.2.1 Community Management

The first issue is the quality of the management personnel. For the convenience of management, a simplified approach has been adopted, whereby management personnel of ecological migrant settlements is principally appointed by those who relocate, and thus, there is an intimate geopolitical and acquaintance relationship between the community management personnel and ecological migrants. As a result, since the quality of the residents in the original community is generally low, the quality of the management personnel elected from them is usually low as well.

The second issue concerns the management methods. Since the communities run by the management officials are acquaintance societies, most of the problems that arise are treated with a blind eye or even indulged and condoned, making them more difficult to solve. For example, in the areas of family planning, fighting for poverty subsidies for residents in violation of the law, and the transaction of homestead and land by community residents, not only have they failed to fulfill their due responsibilities as management personnel, but they have also fostered undesirable practices in ecological migrant settlements, which are harmful to the healthy, harmonious, and orderly development of the entire community.

4.2.2.2 Utilization of Infrastructure

At the same time, despite the establishment of high standard school facilities and medical equipment along with the supplement of a certain number of service personnel in the ecological migrant settlements, it is difficult to achieve the desired effect due to migrants' concern over quality.

The first is the problem of the use of teaching facilities. Improving the cultural quality of the entire population is an enduring and difficult task. The vigorous development of education in ecological migrant areas is an imperative initiative to ensure the sustainable development in such areas. However, as a result of the growing importance people attach to the education of their children, of which ecological migrant areas are no exception, many of them volunteer to take their children outside their hometown for education. This approach is beneficial for the ecological migrants themselves, as it solves both the problem of their children's education and provides them a way out of leaving home for work; however, it is detrimental from the perspective of the ecological migrant settlements. The reduction in the number of children affects not only the motivation of the education practitioners and the quality of the teaching; meanwhile, the complete teaching facilities will fail to function properly over time. As time goes on, school facilities with a high starting point will be abandoned, leading to a new difficulty in accessing schools. It is adverse not only to the development of ecological migrant settlements, but also the improvement of the overall quality of the nation.

The second issue relates to the use of medical facilities. Ensuring the physical and psychological well-being of the people is an important element in building a moderately prosperous society in all respects, and the medical service branches established in the migrant settlements, including full-time staff and corresponding equipment and facilities, have been well received and praised by the community residents. However, such service branches are after all only a rural community healthcare institution, so its service contents are relatively inferior and incomplete. In particular, as people's living standard improves and transportation becomes more convenient, patients tend to be served by higher level of healthcare institutions. In the meantime, health service branches are remunerated based on the number of persons served. Therefore, the income of institutional service staff is significantly impacted. On the one hand, there is an urgent need to improve the service level of medical staff; on the other hand, the number of service objects continues to decrease. How to address this issue has become key to stabilizing medical workers and, of course, to improving the health of people in ecological migrant settlements.

4.2.3 *Self-Initiated Migration Issue Becomes Acute, Speculation is on the Rise*

The so-called self-initiated migration refers to the efforts by desperate poor rural residents to relocate using whatever means at their disposal after having been left out of government-sponsored migration programs. Self-initiated migration comes along with government-sponsored migration programs. As ecological migration proceeds, the issue of self-initiated migration exhibits two phases.

4.2.3.1 The First Phase is the Emergence of Self-Initiated Migration

The main manifestation of this is that, because of the lack of access to the preferential policy of government-sponsored migration programs and a strong desire to be free from destitution, self-initiated migration is adopted, using all their money to purchase the land and houses of the migrants with weak minds or speculators among the enrollees in government-sponsored migration programs, so as to fulfill their own wishes. There is a rational side to the emergence and existence of this phenomenon and also an irrational side to it. On the one hand, self-initiated migration compensates for the labor shortage in ecological migrant settlements and ensures the normal development of existing communities; on the other hand, the illegal self-initiated migration status creates a series of social problems and their own problems, which in turn affects the normal development of ecological migrant settlements to some extent.

4.2.3.2 The Second Phase is When Self-Initiated Migrants Seek Benefits

This is shown in that during the 12th Five-Year Plan period, some of the households that had already voluntarily migrated were included due to the whole-village relocation approach of ecological migration. In order to benefit from the state policies, self-initiated migrants submit their expense that should be paid to the State and thus occupy the homestead and land and relocate with everyone to the ecological migrant settlements; however, since the development land of self-initiated migrants is clearly superior to the current occupied land, the occupied land is sold or leased in order to obtain the corresponding return and reward. The crux of the problem is that they occupy rather than live there, which in turn influences the settlement rate in the community. Although the People's Government of Ningxia Hui Autonomous Region has introduced policies to address self-initiated migration, the problem is still prominent due to low compensation and weak implementation of the policies. While the old self-initiated migration problems remain unsolved, the new ones keep growing.

Ecological migration is a project that benefits the people and is also a huge and complex systemic project. Therefore, some loopholes in the selection of ecological migration subjects are inevitable. In the process of relocating migrants, despite the strict monitoring and supervision mechanisms, some speculators exploit their personal connections or rights to sell houses and land indexes, obtain ownership of houses, or sell the indexes they have for profit, or they may choose second relocation because they are dissatisfied with the development environment of their host place, resulting in unoccupied houses registered in their names.

During specific research activities, we saw advertisements for "houses for rent" and "houses for sale" in the labor migrant settlement of Guyuan City, indicating that this phenomenon has become public. At the same time, in Hongsipu District, we also saw wide-open doors and sealed windows, which according to the entourage, belong to migratory bird type migrants, that is, when the Government comes to inspect or

Table 4.27 Renting and sales of houses in Hongde Village of Hongsipu (%)

	Yes	No	NA
Renting	16.30	53.10	30.60
Sales	18.40	38.80	42.80

offers concessions, they can be recruited by a single phone call, and once they are finished, they will return to their place of origin or somewhere else.

Table 4.27 shows that nearly 20% of the respondents believed that the phenomenon of renting or sales existed. These phenomena are typical speculative behaviors, which will harm the development of ecological migrant settlements and not only seriously violates the "no-buy or sell of homestead and land" rule of these communities, but also has a negative demonstration effect on other established settlers, leading to a low settlement rate in ecological migrant settlements.

4.2.4 Low Quality and Weak Social Adaptability Confront the Migrants

People's high or low overall quality is critical to their own development and that of the community. The research results on ecological migrant settlements over the years indicate that the cultural quality of ecological migrants is generally low. This research presents the same result as well.

For ecological migrants, cultural quality is positively correlated with development capability. As can be seen from Table 4.28, the cultural quality of ecological migrants is relatively low, which in turn leads to their relatively low overall quality.

Table 4.28 Education level of the respondents (%, person)

	Xiji County		Yuanzhou District		Yuanzhou District		Hongsipu District	
	Ecological migrants		Labor migrant		Ecological migrants		Mixed ecological migrants	
Illiterate	9	17.3	23	46.0	27	48.2	21	42.9
Primary school	11	21.2	19	38.0	21	37.5	13	26.5
Middle school	24	46.2	6	12.0	8	14.3	12	24.5
High school	5	9.6	2	4.0	0	0.0	3	6.1
Junior college or above	3	5.8	0	0.0	0	0.0	0	0.0

4.2.4.1 Poor Adaptability to the New Development Environment Due to Low Cultural Quality

The overwhelming majority of the ecological migrants come from the backward southern mountainous regions. They have low cultural quality, conservative mind, weak legal awareness, and difficulty in changing their mindset; meanwhile, they lack professional skills, face difficult employment, and have weak innovative capability and limited enthusiasm for entrepreneurial projects. Therefore, they fear that they are unable to adapt to new jobs and that they will even fail to find themselves a new job.

4.2.4.2 Persistent "Wait, Depend and Demand" Mentality Due to Low Overall Quality

Ecological migrants from the southern mountainous regions of Ningxia have long lived in the rural villages and maintained a traditional farming lifestyle, and the Government has been adopting a transfusion type of poverty alleviation there, that is alleviating poverty through subsidies and relieves. Therefore, though it has mitigated or improved the living conditions of the impoverished population to a certain extent, it has also led to the formation of a "wait, depend and demand" mentality among some people. Despite the fundamental change in the lifestyle and production of ecological migrants, a few people have difficulty in changing their mindset and they would rather sit at home and wait for subsidies than work. The idea of "wait, depend and demand" still exists. Meanwhile, in the face of rising prices and growing living expenditure, as long as there is a lack of an effective source of income and frustration with the poor development environment, they will selectively withdraw from ecological migration.

4.2.4.3 Inability to Comprehend and Implement Policy Due to Low Overall Quality

It should be said that the vast majority of ecological migrants are aware of the ecological migration policy, but not many of them truly understand its specific connotation, let alone use it to protect their rights and interests and solve their own problems. They often understand only part of the problem. They either take the policy out of context, which leads to unnecessary group petitions and seriously affects the normal development of ecological migrant settlements, or they wait passively because of their lack of understanding on specific connotation of the policy, resulting in their inability to effectively protect their legitimate rights and interests.

4.2.5 Migration Policy Needs to Be Improved, with More Attention Given to Special Groups

Policy is the lifeblood of concrete work, which is also the case with ecological migration. Of course, as time goes on and the development environment changes, policies should be improved and refined accordingly, especially with regard to special groups.

4.2.5.1 With Regard to Poor Households

According to the ecological migration policy, relocated households are required to pay a certain amount of fee before relocation, and poor households have difficulty in doing so due to a lack of funds.

4.2.5.2 With Regard to Multi-Person Households

According to the 12th Five-Year Plan ecological migration policy, all household splits before December 31, 2009, will be recognized and none after that will be acknowledged. As a result of the implementation, an ecological migrant house of 54 m^2 is unable to properly accommodate more than six people of three generations living a normal life, resulting in the emergence of multi-person household problems.

4.2.5.3 With Regard to Single-Person Households

This is one of the most challenging issues to address in the ecological migration process. The single-person households are often forbidden to relocate, which leads to social conflicts.

4.2.5.4 With Regard to Development Environment

Prior to ecological migration, staff have repeatedly stressed that the development environment of the future host place is excellent that once it is found to be different from what have been advertised after relocation, or the original conditions are again stranded for various reasons, the development of the ecological migrant settlement and the sustainable development of the ecological migration will be affected.

Chapter 5
Social Inclusion of Ningxia Ecological Migrants

Xihong Shu and Jun Nie

In 1982, the State Council decided to initiate a Three West regions agricultural construction project to fundamentally solve the problem of food and clothing for the people in the contiguous destitute areas in Gansu and Ningxia. In September 1983, the Leading Group of the Three West regions agricultural construction further identified the guideline that "if there is water, then implement the project in the water; if there is dryland, implement it on the dryland; and if there is neither water nor dryland, just try otherwise." Based on this poverty alleviation policy, the People's Government of Ningxia Hui Autonomous Region formulated the strategy of "benefiting the Hetao area and alleviating the poverty in Xihaigu region," mobilizing people in the southern mountainous regions of Ningxia, where resources were limited and people lived in areas with narrow living spaces and lack of basic living conditions, to move to the more resource-rich wastelands with irrigation conditions for developmental production construction. Since then, Ningxia's poverty alleviation work has shifted from rescue to development, and from transfusion to creation. Faced with the current situation of large impoverished population, rooted poverty, poor living conditions, and challenging development, the People's Government of Ningxia Hui Autonomous Region has, at different times, implemented political migration projects such as the Diaozhuang Migration,[1] the 1236 Project (Yellow River-Pumping Irrigation Project) and ecological migration. A variety of means and flexible methods are applied for the relocation and resettlement of migrants, resulting in a win–win situation in which the migrants have grown from poverty to prosperity while the ecological environment has effectively improved. By 2015, the accumulated number of migrants in Ningxia

X. Shu (✉) · J. Nie
College of Ethnology, Northern Minzu University, Yinchuan, China
e-mail: shuxihong@163.com

[1] Diaozhuang, which is also known as Ladiaozhuang, is a type of agricultural business in the Northwest Loess Plateau during the late Qing and Ming dynasties. Farmers settle in one village while cultivate and farm in another, so they have to travel back and forth between the two places. One village connects two places, so it is called "Diaozhuang."

X. Wang (ed.), *Ecological Migration and Targeted Poverty Alleviation in Ningxia*, International Research on Poverty Reduction,
https://doi.org/10.1007/978-981-19-7888-3_5

had reached 1.16 million, accounting for one-sixth of Ningxia's population.[2] The per capita disposable income of Ningxia ecological migrants amounted to 5638 yuan,[3] and the income of a few rich ones had surpassed that of rural residents in Ningxia. Per capita disposable income has increased more than tenfold compared to the initial period of the migration. The remarkable achievement of the development of Ningxia ecological migration proves that ecological migration is a strategic initiative to solve the fundamental poverty problem in Ningxia, and is of great significance to speeding up the process of poverty alleviation, transforming development models, improving people's livelihood, coordinating regional development, and building a moderately prosperous society in all respects. However, as a difficult and complex social system project, the ecological migration project also faces many problems in its implementation. Migration not only brings changes in the location of the productive and living spaces, but also leads to changes in the migrants' productive lifestyle, social relation structure, customs, identity, etc. In the face of these changes, most migrants will experience feelings of discomfort, isolation, and exclusion within a short period of time, making it difficult for them to establish themselves in the host place and develop the economy with confidence. Therefore, certain methods should be applied to help ecological migrants to adapt to the natural and social environment of the host place, reduce the probability of their return or second relocation, enable them to integrate into the local social life and community environment as soon as possible, and ensure their stability after relocation. Only in this way will migrants be able to devote themselves to production so that the goal of entering a moderately prosperous society in all respects with the whole country by 2020 can be achieved. To this end, this chapter delineates the dimensions of social inclusion of ecological migrants based on questionnaires and interviews, describes the current situation of their social inclusion, explores the salient factors affecting it, and proposes possible paths to promote it.

5.1 Definitions and Current Conditions

5.1.1 Defining Social Inclusion

As the starting points and perspectives of studies vary, so do the researchers' understandings and definitions of social inclusion. Some researchers view migrants as a disadvantaged group and consider social exclusion as an objective existence in their social life. On the premise of eliminating social inequalities, social inclusion stands for actively promoting opportunities for participation, whether at work, in education or at the broader social level.[4] According to Liu Jian'e, social inclusion is the

[2] Yang (2015).

[3] The data are from the Statistics Bureau of the Ningxia Hui Autonomous Region.

[4] ESFP (European Structural Funds Program) "Further Information: Social Inclusion" 2007–2013 European Structural Funds Program.

inclusion of individuals and groups in a given society into the mainstream society in the process of equal involvement by gaining access to basic economic and political rights and a broad range of social rights, including rights to employment, democratic elections, policy decisions, social services, and urban cultural life, through structural adjustment and self-adaption.[5] This definition of social inclusion is based on the social exclusion theory and the two prerequisites of an equal living environment and a sense of active participation. In the European and American societies, the concept of social inclusion originates from social integration but has gradually replaced it. Park and Burgess consider social integration to be "a process of interpenetration and fusion in which persons and groups acquire the memories, sentiments and attitudes of other persons or groups, and, by sharing their experience and history, are incorporated with them in a common cultural life." According to Chinese researcher Yang Congmin, "social integration is the fusion and interpenetration of the original culture of the migrants with the culture of the host place to form a social cultural system with something new.[6] It can be seen that social inclusion based on the social integration theory is more of a cultural integration with each other and emphasizes a state of integration with equal goals. Chen Chengwen et al. argue that social inclusion is a process of social action in which disadvantaged subjects interact reflectively and continuously with individuals and groups in a given community.[7] Social inclusion is essentially a social action that is dynamic, reflective, continuous, and interactive.

In summary, we assume that the concept of social inclusion of ecological migrants should be: It is a process in which ecological migrants regain social identity and respect from others, and obtain a sense of dignity and satisfaction in life by actively accessing and making rational use of the social resources in the host place, maintaining social relations through continuous social actions, and reflecting on and adjusting their own behaviors and social relations during the actions on their own initiatives based on the equal environment created by the government.

5.1.2 Status Analysis of the Social Inclusion of Ecological Migrants

5.1.2.1 Economic Life Inclusion

Restricted by their own conditions, migrants may face problems including discrimination and low income in the process of economic life inclusion. Regarding employment, 61.69% of the migrants interviewed said that they "have never been discriminated by the local people in job-seeking because of their migration status." However, 53.11% of the respondents stated that "it is difficult to find a job after relocation

---

[5] Liu (2010).

[6] Yang (2010).

[7] Chen and Sun (2012).

and the opportunity is extremely limited." This shows that the employment environment in the host place is relatively equal and migrants have access to equal employment opportunity, but employment opportunities and channels are scarce. In terms of income, in 2015, migrants with annual income of less than $100, $100–500, $50–900, and $900 or more accounted for 22.01%, 20.02%, 12.20%, and 45.77% of the interviewees, respectively. The per capita disposable income of rural permanent residents in Ningxia in 2015 was 9119 yuan.[8] A little over half, or 54.23% of the migrants interviewed had an annual income lower than the average of the autonomous region. The gap is quite large; some of the migrants are economically disadvantaged. In terms of vocational training, only 21.34% of the migrants surveyed said that they have "attended various types of government-organized labor skill training sessions," of which 41.38% indicated that such training "don't help at all or don't help much." Skill training is directly linked to employment, and weak vocational skill training is unfavorable to the migrants' inclusion into the local economy.

5.1.2.2 Social Behavior Inclusion

In terms of life interactions, 30.47% of the migrants interviewed said that they "often visit other migrants or local residents"; 84.83% of the respondents said that they "have made friends with other migrants or locals." This shows that the migrants interact quite frequently and deeply among themselves and with the local residents. This is due to the reconfiguration and improvement of the social networks. As for participation in social activities, 65.67% of the migrants interviewed responded that they "often or sometimes attend the weddings and funerals of other migrants or local residents," and the majority of the migrants maintain good relations with others through attending weddings and funerals or other social activities. In terms of participation in community activities, 25.27% of the migrants surveyed said that they "have never participated in group activities organized by the village committee," of which 12.5% explained that the reason was they "do not want to have too much contact with people." This suggests that migrant's participation in community life is generally low and a few of them are unwilling to actively take part in community activities and interact with others. In terms of participation in political life, 44.03% of the migrants surveyed checked "Never" to the question "Have you ever voted in the village committee elections in your current place of residence," and 35.20% of them replied "Yes, but it is required by the village officials"; only 20.77% of the respondents said that they "have actively participated in the elections." The low level of political participation of migrants is evident.

[8] The Statistics Bureau of the Ningxia Hui Autonomous Region. "The Statistical Bulletins of the National Economic and Social Developments of Ningxia Hui Autonomous Region in 2015." *CNStats*, May 25, 2015.

5.1.2.3 Culture Acceptance

In terms of languages, since both Hui and Han people use Chinese characters, there is no language barrier between the Hui and Han migrants and between the migrants and the local residents except dialect difference. 80.72% of the migrants interviewed said they "can almost understand all dialects of the other migrants and the local residents"; most of the migrants are used to communicating with others using the dialect of their hometown in their daily communication. As to "Would you like to learn the dialect of other migrants or locals," 61.32% of the migrants interviewed answered "Yes, as it would facilitate their communication with others." This shows that most of the migrants are open to communication. In terms of the festive customs, 67.04% of the migrants interviewed said that they "have never participated in any festivity unique to other migrants or the local residents"; 62.56% of the migrants surveyed (Hui) replied that they "will not absorb or learn from the wedding and funeral customs of the Han migrants or the local residents." This suggests that migration has not exerted much impact on the migrants' festive customs, but choose to maintain their traditional festivals as well as wedding and funeral customs; meanwhile, they are reluctant to actively absorb and learn other strongly heterogeneous ethnic and regional cultures. In-depth cultural exchanges between Hui and Han migrants and between them and the local population are uncommon. Both Hui and Han cultures are important multicultural elements of China. The subconscious cultural perception deviations in some of the migrants have both prevented them from assimilating to or learning from the best cultures of other ethnic groups and hampered exchanges among the migrants themselves.

5.1.2.4 Psychological Inclusion

In terms of feelings and interactions, 15.67% of the migrants surveyed said that they "often or occasionally experience anxiety in their interaction with the local population because of their migration status." It can be learned that because of their special and disadvantaged status, a small number of migrants will develop a sense of inferiority and self-protection, which may lead to their unwillingness to contact with other migrants or local residents. As for social identity in their place of residence, 47.51% of the respondents said that the "have deeper feelings toward their place of origin," 35.07% of them indicated that they "have deeper feelings toward their current place of residence," and 17.42% said that "it is hard to say." On the one hand, it shows that a proportion of migrants still have a deep attachment to their hometown after relocation; on the other hand, it illustrates that some migrants have not adapted well to the life in their current place of residence and have a low level of social identity to it. In terms of identity, 84.33% of the migrants interviewed preferred to be a local, and in general, the longer they have been relocated, the more pronounced the tendency to be a local. A further 94.03% of the migrants interviewed wanted to be considered as locals by the others. It shows that given the many social resources that come with being a local, while part of the migrants have not considered themselves as one, they still want to be recognized as one by the others. As for life satisfaction, a few of the migrants interviewed checked "Great dissatisfaction or limited satisfaction with life in their current place of residence," highlighted by dissatisfaction with one's economic income, which accounted for 54.48% of the respondents. Dissatisfaction with life in the current place of residence increases the risk of second relocation, nor is it beneficial to the psychological inclusion of migrants.

5.2 Factors Affecting the Social Inclusion of Ecological Migrants

The factors affecting the social inclusion of Ningxia ecological migrants include both external and internal factors. The external factors are mainly objective social environmental factors; the internal factors mostly relate to the intrinsic characteristics of individual migrants.

5.2.1 *Cultural Differences*

"The Hui culture is a multiplex culture gradually formed by the collision, dialogue and adaption of Islamic and Confucian civilizations in China. It has many similarities with the Han culture which is based on Confucianism and Taoism."[9] Even so, the Hui and Han cultures are still highly heterogeneous, especially in terms of food, marriage, religious beliefs, and customs, and the differences between the two cultures are significant. The relocation has reconstructed the settlement pattern of the host place, and the degree of mixed settlement of Hui and Han migrants has increased, which to some extent provides more opportunities for interaction and exchange of the migrants of these two ethnicities. Although there is a growing need for Hui and Han intercultural exchanges in daily production and life, their cultural exchange still stays at the superficial level. Especially for those migrants who have little contact with other ethnic groups prior to relocation, frequent contact with other ethnic groups in different settings such as the place of residence, workplace, schools can easily cause discomfort, while superficial perceptions of other cultures and the discomfort during the interaction will further weaken their willingness to interact with other ethnic groups.

[9] Ma (2001).

5.2.2 Economic Deprivation

Economic life inclusion is a process in which individual migrants gain access to equal rights and opportunities to employment, social security, and opportunities with the majority of migrants or indigenous residents in their place of residence and get rid of poverty with a limited financial distance from others. Data indicate that 41.67% of the migrants interviewed indicated that their greatest difficulty in integrating into the local community was economic hardship, which is highlighted by the significant imbalance in the family's income and expenditure after relocation. Prior to relocation, most migrants rely on traditional agriculture for livelihood, and most of their farm produces are low-value grain crops, with low agricultural returns and household savings. After relocation, the government offer migrants some preferential policy and financial support. However, due to the limited financial resource, it can only relieve the economic pressure on migrants to a certain degree. In the case of migrant housing during the 12th Five-Year Plan period, for example, each migrant household still had to bear a small part of the construction cost of their houses, which was about 150,000 yuan. This was still a significant expense for poor migrant families. In some of the migrant villages relocated during the 12th Five-Year Plan period, all lands have been transferred and the migrant's main livelihood is nonfarm employment. However, in recent years, as China's economic development enters a new normal, economic growth slows down and some industries closely related to migrants' livelihoods, such as real estate, enter recession, the migrants' livelihoods are seriously affected, and some migrant families are left in economic difficulties. For example, in 2015, 29.73% of the migrants interviewed indicated that "their household expenditure has exceeded income, leading to indebtedness"; quite a large proportion of the migrant families cannot earn enough to cover their expenditure. The economic deprivation has become one of the largest barriers hindering migrants' inclusion to the local society.

5.2.3 Discrimination and Prejudice

The social inclusion of migrants must not only emphasize the initiative of individual migrant, but also take into account the objectivity of the external environment and the degree of acceptance of others. An equal and just social environment along with an open and inclusive social mentality can better facilitate the social inclusion of migrants. One of the difficulties for migrants to integrate into the society of the resettlement areas is lack of recognition and respect, which is due to the discrimination and prejudice that migrants experience from others. This discrimination and prejudice mainly come from two aspects. From the cultural point of view, the biased perception Hui and Han people have on the cultural differences of each other often leads to ethnic discrimination and prejudice. Moreover, this belief in the superiority of one's own culture or ethnicity over the others is often passed down from generation to generation until it becomes a hardly detectable traditional and customary perception. Meanwhile, it is accompanied by strong emotional factors and behavioral tendencies, implicitly influencing the development of ethnic relations. From an economic point of view, the constant influx of migrants, while booming the local economy, also occupies a significant amount of social resources. Believing that migrants have narrowed their living spaces, some indigenous residents grow discontent, which makes it difficult for them to view migrants objectively and impartially. Their negative psychological and emotional impact cannot be eliminated without the adoption of positive strategies to accommodate the inclusion of migrants.

5.2.4 Lack of Social Support

After relocation, "migrants in the new cultural context experience feelings of loss and helplessness as a result of being disconnected from their familiar cultural environment, or rejected by the people in the new cultural context. At the same time, they feel frustrated and defeated as a result of being unable to position themselves appropriately in an unfamiliar social network. Without adequate social support, similar psychological perceptions of migrants often lead them to vent their negative emotions in overly aggressive ways, seriously threatening the harmony and stability of the resettlement area."[10] According to data, 77.61% of the migrants interviewed indicated that they would seek help from relatives when they encountered difficulties, followed by friends of same ethnic groups (65.17%) and neighbors (33.08%). A sense of kinship by blood and a sense of belonging by national sentiment offer migrants in the initial stages of relocation indispensable psychological support and emotional solace. In the early stages of relocation, migrants' personal social networks are disrupted to a certain extent, and they lose not only the spiritual solace that comes from kinship, geographical relationship, occupational relationship, and so on, but also the resources that social networks provide. In the "scattered relocation and fill-in resettlement" phase, migrants' personal social networks are most severely disrupted, and their relationships with relatives, friends, neighbors, and colleagues are more thoroughly dismantled, with most migrants only receiving support from their own family, which is far from sufficient for their inclusion into the migration society. During the "whole-village relocation and whole-village resettlement" phase, the social networks of the migrants are better preserved, especially as some of their relatives, friends, and neighbors are also relocated with migrants' spatial location. In the new cultural context, the old acquaintances provide the migrants with some psychological and spiritual support, significantly reducing the negative impact of relocation on them. Migrants lack not only spiritual support, but also material assistance. According to many of the migrants interviewed, what they lack most for economic development at present is production funds. Migrants' own funds are extremely limited, loan thresholds

[10] Feng and Nie (2013).

are relatively high, and the sources of production fund are few, unable to effectively support production or entrepreneurial projects, and the supporting role of the government sectors as well as financial enterprises has not been well played.

5.2.5 *Self-Initiated and Involuntary*

Policy-based migrants distinguish between self-initiated and involuntary migrants, with some notable differences in terms of social inclusion. For self-initiated migrants, they are more positive, open, and tolerant in their communications with others, and willing to absorb and learn from the strengths of others while maintaining some of their own inherent cultural traditions. When caught in conflict with others, they are more rational and more ready to compromise and step back. It is their wish to live in harmony with other migrants and indigenous residents and achieve mutual development. In the meantime, involuntary migrants lack the desire to move to their host place, so they tend to be emotionally pessimistic, negative, repellent to others, reluctant to interact with others, and prone to self-containment and isolation; meanwhile, they normally grow overly emotional in dealing with problems, so they can easily stand opposite with the others, resulting in conflicts. Therefore, it is imperative to develop thorough policies and measures before relocation to promote self-initiated relocation of migrants so that they can integrate into the local society with more positive attitudes.

5.2.6 *Lack of Human Capital*

Lack of human capital is another important factor that contributes to the difficulties of migrants' social inclusion. Human capital is a specific type of capital in which migrants accumulate knowledge and improve skills through certain education and training, which enables them earn wages and other benefits. The data show that among the migrants interviewed, 42.29% of them are illiterate or semi-literate, 22.26% with primary school education, 23.38% junior high school, and 12.07% graduated from high school and above. It can be seen that migrants are a group of people with a generally low level of education. This means that most migrants lack knowledge accumulation, which prevents them from acquiring new vocational skills. As a result, they are unable to find better-paid jobs or integrate into the local economic life in a short time. To address the problems of migrants' low level of education and poor vocational skills, the government should organize more labor skill training sessions to promote the upgrading of migrants' human capital. However, due to the short duration of the training, its low pertinence, and narrow coverage, the implementation brings no satisfactory result. In addition, though children of migrants generally take compulsory education, they still face difficulties in accessing higher levels of education. Some migrant families have limited wealth accumulation so they cannot afford the high education cost, leading to them becoming poor as a result of debt brought by the education of their children.

5.2.7 Years of Residence in Current Place of Residence

Western scholars have proposed many cultural adaption theories from an individual perspective. Oberg has summarized the process of cultural adaptors receiving culture shock into four stages, namely the honeymoon, crises, recovery, and adjustment stages. The U-Curve Hypothesis developed by Lysgaard states that individuals enter a new cultural context with novelty, enthusiasm, and a high level of satisfaction, followed by a period of loss or disappointment due to maladjustment and they will remain at a low level of satisfaction, and finally, they will experience a gradual recovery to the highest level of satisfaction. In addition, there is also the W-Curve Hypothesis of Gullahorn and Adler's Five Stage Model of Cultural Adaption, etc. No matter how these theories develop, they all have one thing in common, that is the time line. As the migrants spend more time in the host place, they become more accustomed to the different culture there. In general, when migrants are in a new social situation, they are bound to encounter difficulties at some point of the initial stage, which may lead to a certain sense of loss and helplessness, and a decrease in their satisfaction with their surroundings. With longer periods of residence, along with resolution of problems in life and adaption to the surrounding environment, the migrants will gradually grow out of the period of loss, their life satisfaction will increase, and they will be better integrated into the local social life. However, this process is not granted either, but presupposes a gradual improvement in the situation of the migrants themselves. Investigation finds that a small number of the migrants fail to integrate well as they live longer in the host place and that they eventually chose to return or relocate for the second time, which are obvious manifestations of the failure of migrant inclusion into the society.

5.2.8 Willingness and Ability to Interact

The willingness to interact is one of the prerequisites for migrants' social inclusion. After relocation, the economic and social status of some migrants declines, and the disparity before and after relocation makes them vulnerable to feelings of inferiority, accompanied by behaviors of self-isolation and excluding others. With the lack of recognition from indigenous population and other migrants, the inferiority feeling of a small number of migrants increases, which makes it harder for their psychological inclusion. Moreover, even if there is a willingness to interact, it does not mean that they can smoothly interact with the others, which also depends on the interpersonal skills of the migrants. Interpersonal skills include the ability to express and comprehend interpersonal integration and problem-solving ability, among which the ability to express and comprehend is related to the degree of migrants' social adaption, while interpersonal integration ability is relevant to the degree of their social inclusion. Although the ecological migrants come from different counties in central and southern Ningxia, generally they have retained comparatively strong consistency in their languages and behaviors, and there is no major problem in their mutual understanding of each other's expressions. The degree of migrants' mutual adaption is quite high. The interpersonal integration ability determines the migrant's ability to develop and maintain the relationships with others, and this ability depends on whether the migrants can treat others sincerely in daily interactions, which is highly related to personal attitudes.

5.2.9 Age Difference

The social inclusion of migrants varies considerably by age. The survey finds that compared with middle-aged or elderly migrants, young migrants take shorter time to integrate into the society with a deeper degree of inclusion. Young migrants are full with desires and needs of the outside world. Meanwhile, middle-aged and elderly migrants are physically unfit, with lower ability to learn and accept new things and stronger attachment to their hometown, leading to their lack of competitiveness to young migrants. They need longer time to adapt to unfamiliar social environment and are more difficult to integrate into it.

5.3 Social Inclusion of Ecological Migrants and Possible Paths

5.3.1 Face National and Cultural Difference Seek Common Ground While Preserving Differences of Hui and Han People

First of all, active guidance should be offered to Hui and Han migrants on facing the national and cultural differences as well as personalities, which is to consolidate the cultural basis for social inclusion of Hui migrants. Under the premise of equality and mutual benefit, we establish the consciousness and understand the feature of pluralistic integration of Chinese culture, and recognize and respect the cultural differences and national characteristics of the Hui and Han people in terms of their historical traditions, lifestyles, customs, value, and so on that are formed in the course of development and evolution, so as to create a positive atmosphere of mutual understanding, respect, and tolerance among Hui and Han migrants. Secondly, it is necessary to fully explore the modern political concepts contained in the traditional Hui culture, especially the rich modern political civilization concepts embodied in the Islamic doctrine, such as the concepts of harmony, equality, fairness, justice, patriotism, and so on. Through the exploration and promotion of these modern political civilization concepts, Han migrants' knowledge on the traditional Hui culture can be deepened. Finally, the core values of socialism should be used as a guide to coalesce the common culture and value cognition between the Hui and Han people, strengthen their shared recognition of Chinese culture, and thereby reinforce the close ties between Hui and Han migrants where "they jointly receive rewards and face challenges," and create stable national psychological conditions for the social inclusion of Hui migrants.

5.3.2 Promote Follow-Up Industries, Strengthen Precision-Targeting Poverty Alleviation

In terms of follow-up industry development, various types of agriculture-related funds can be integrated to improve the lands unsuitable for cultivation, or large-scale land transfer can be facilitated to enterprises, in order to enhance the efficiency of farmland output; migrant villages are encouraged to introduce new breeding models such as the "government + company + cooperative + farmers" model, forming an interest-linked mechanism of mutual sharing of benefits and risks to protect the interests of farmers to the fullest extent; the role of the market in allocating labor resources will be given full play, and market carriers such as labor brokers and labor intermediaries are encouraged to develop. Meanwhile, labor output is promoted through multiple channels according to the characteristics of the migrants, enhancing the targeting of migrant output and stabilizing the service income of migrant workers. In the meantime, enterprises are encouraged to enter villages and establish types of enterprises such as featured agricultural processing companies and family workshops in the migrant villages to solve the employment issues of women as well as middle-aged and elderly workers. Regarding precision-targeting poverty alleviation, a multi-level assessment system has been adopted to improve the procedure for identifying poverty alleviation targets; comprehensively taking into consideration of the assets and circumstances of migrant families, the accuracy of the assessment of poverty alleviation targets is enhanced through investigations on their income, assets, and household circumstances; the causes of poverty in migrant families are accurately identified, and poverty alleviation measures are taken in accordance with different causes, such as illness, education, disability, and loss of workforce. By the development of follow-up industries and precision-targeting poverty alleviation measures, the income of migrants will be raised while their expenditure will be reduced so as to balance the two, resolve their economic difficulties, and lay a solid economic foundation for social inclusion.

5.3.3 Deepen the Mutual Cognition Between the Hui and Han People, Strive to Eliminate Discrimination and Prejudice

Efforts should be made to deepen the knowledge and understanding of the Hui and Han migrants on political, economic, cultural, and social aspects of each other, creating a social environment of equality, solidarity, mutual support and love, and eliminating discrimination and prejudice among ethnic groups at root. On the political front, the villager autonomy work in migrant villages is comprehensively promoted in accordance with the law, and the construction of the two village committees has been strengthened. The training of officials of the Hui ethnicity is reinforced, and the proportion of the Hui officials is increased through greater financial and human resources support for migrant villages, so that the Hui migrants can fully participate in the democratic construction of villages and obtain equal rights and status. At the same time, emphasis is placed on improving the ability of officials on handling conflicts and disputes among the Hui and Han people, resolutely "rectifying and eliminating words and behaviors of discrimination or discrimination in disguised forms that are against ethnic minorities or injure national feelings," and safeguarding the unity of the Hui and Han people.[11] In the economic aspect, social resources are allocated equitably and economic development of the Hui migrants is promoted according to their own advantages. In the meantime, a complementary and mutually beneficial economic structure is gradually formed in the common economic development of the Hui and Han migrants. The gap in economic development between ethnic groups is narrowed, and the Hui and Han migrants are enabled to achieve common wealth. In terms of culture, cultural exchanges among the Hui and Han migrants and the indigenous population are promoted through the organization of cultural and recreational activities that are popularly welcomed by the migrants on major festivals and traditional national holidays. Deeper awareness and understanding are established through positive interactions, and thus, the cohesion of migrant settlements is enhanced. In the social aspect, education on national unity is strengthened, and media and channels are exploited to firmly grasp the theme of "achieving common prosperity and development through concerted efforts" for the Hui and Han people, promote the Party's and the State's national theory, national policy, national laws and regulations, and basic national knowledge propaganda, and create a favorable atmosphere for the harmonious living, cooperation, and peaceful development of the Hui and Han migrants, and those of migrants and indigenous residents.

[11] Xi (2014).

5.3.4 Restructure Social Relation Network and Offer Necessary Social Support

In terms of spiritual support, as for the Hui migrants, mosques can be rationally planned and built according to the religious affiliation of the Hui migrants in the village, the size of the population, and their different original regions. This is to reconstruct the internal ties among the Hui migrants with religious beliefs at its core, so that they can gradually receive spiritual support from their own ethnic groups. In addition, migrants' sense of loss and helplessness is alleviated and they are helped through mental difficulties by the stimulation of a sense of shared community among Hui, Han migrants and the indigenous population, the creation of an equal interpersonal network, the exchange of resources, information and mutual assistance among members, and the gradual establishment of a social resource allocation system that simultaneously emphasizes kinship, geographical relationship and occupational relationship".[12] As for material support, the government has been actively expanding access to financing for migrants and encouraging the establishment of microcredit companies for ecological migrants, with emphasis on supporting migrants' farming, breeding, and other entrepreneurial projects. At the same time, banks in resettlement areas are encouraged to innovate their lending models, introduce new products such as migrant loans, prioritize loans for migrants with financing difficulties, provide preferential interest rates, relax loan guarantee conditions, and effectively address the shortage of productive capital for migrants.

5.3.5 Innovate Education Incentive Mechanism and Enhance Cultural Skills of Migrants

A direct subsidy mechanism for vocational skill training is established to straightly subsidize the funds of vocational skill training to individual migrants who, according to their availability and willingness, choose the training institutions tendered by the government to participate in professional training. Relevant government departments will assess, register, and record the effectiveness of their training. The training funds for Hui migrants who are directly trained by enterprises according to their needs are straightly subsidized to the enterprises where they are employed. At the same time, individual migrants that have obtained national vocational qualifications through vocational skill training may receive another incentive allowance from the government. An incentive mechanism for higher education is established. For migrants' children who continue in higher education beyond the compulsory stage, they will be offered an incentive allowance funded by the government, village collectives, charitable enterprises, and individuals, with a tiered amount depending on their level

[12] Wei (2013).

of education. Through the educational incentive mechanism, migrants are motivated to learn and improve their cultural quality and vocational skills.

References

Chen, C., & Sun, J. (2012). Social inclusion: Sociological significance of a concept. *Journal of Social Science of Hunan Normal University, 6*.

Feng, X., & Nie, J. (2013). Cultural adaptation of Hui ecological migrants in Ningxia with multidimensional model. *Journal of Jishou University, 6*.

Liu, J. (2010). A view on social inclusion for rural-urban migrants in urban of China from the perspective social inclusion policy of European Union. *Urban Studies, 11*, 106–112.

Ma, Z. (2001). On the cultural adaptation in the Hui nationality. *Researches on the Hui Nationality, 3*.

Wei, Z. (2013). A probe into the social integration path of Tibetan ecological migration. *China Tibetology, 1*, 125.

Xi, J. (2014, September 30). Specify objectives and improve the national work ability. *Jiefang Daily*.

Yang, C. (2010). *Equal rights and social integration of migrant workers* (p. 96). Zhejiang Gongshang University Press.

Yang, G. (2015, December 22). Targeted poverty alleviation in Ningxia through ecological migration and the joint efforts of the society. *Economic Daily*.

Chapter 6
Endowment Security Innovations for Ecological Migrants in Ningxia

Lixia Kong and Jun Nie

Most of the ecological migrants were relocated in units of families, and relocations of families with two or more generations are very common. Many elderly people moved from the southern part of Ningxia to the central north area with their children, hoping that their children would be better able to provide for them if their own economic conditions improve. Yet our survey has revealed that for some of the elderly migrants, not only their endowment conditions have not been improved after relocation, they have also fallen into heaps of endowment predicaments. Only by innovating the personal endowment security mechanism for the migrants is it possible to further resolve the endowment predicaments for the ecological migrants.

6.1 Endowment Predicaments for Ecological Migrants

The interconnection of endowment insurance formalities represents one of the major contents of this great people-benefiting endeavor. Since the first batch of ecological migration, interconnection of the migrants' endowment insurance has been implemented between the governments of the host and original places. As of the date of the survey, the endowment insurance of the ecological migrants had been basically transferred to the government of the host place; the ecological migrants were enjoying the endowment insurance benefits for the local urban and rural residents. The host places are generally more economically developed river areas, providing the migrants with

L. Kong (✉)
Northern Minzu University, Yinchuan, China
e-mail: zy1965@163.com

J. Nie
College of Ethnology, Northern Minzu University, Yinchuan, China

X. Wang (ed.), *Ecological Migration and Targeted Poverty Alleviation in Ningxia*, International Research on Poverty Reduction,
https://doi.org/10.1007/978-981-19-7888-3_6

higher endowment insurance benefits. Unfortunately, improved endowment insurance benefits for the migrants after relocation is still less than enough to meet the various daily living needs of the elderly migrants.

In terms of income, in addition to the major source of income of endowment insurance, elderly migrants can also obtain endowment security through agricultural or animal husbandry production, working outside, or support from their children.

Land is an important resource for farmers to maintain subsistence. As China's grain output increases year by year, the prices of grain crops have continued to fall. The agricultural income of the migrants is very trivial. According to the survey, in 2016, the net return on growing corn per mu in the host area was only a few hundred yuan. Relative to the young- and middle-aged migrants, elderly migrants are partly or totally incapacitated. Over time, it becomes even more difficult for elderly migrants to earn income through agricultural production. As data indicate, in 2015, only 32.5% of the elderly migrants had agricultural income; most of them had either left their land uncultivated or circulated it to companies for scale utilization. Among the 10 migrant villages interviewed, four had totally circulated their land. Qihai Village of Zhongning County's Dazhanchang Town had signed an eight-year circulation contract with a company, under which each of the villagers would receive a circulation fee at 600 yuan per mu of land per year over the first four years and 800 yuan per mu of land over the next four years. In Xinghai Village of Zhongwei City's Xuanhe Town, each migrant receives 400 yuan per mu of land per year. For the land-lost elderly migrants, the trivial income of a few hundred yuan per year is hardly able to keep them survived. What's worse is that, the signing of a land circulation contract does not necessarily guarantee that the circulation fee is paid to the migrants on time. There are frequent cases where land has been circulated but the migrants are not paid the circulation fee. A migrant surnamed W from Minnnig Town's Yuanlong Village said, "The land of the village has been circulated. Years has passed, but no money has been paid to us. Those old people in the village have little income. The village has organized us to talk to the company, but no result has been obtained yet."

Before relocation, many elderly migrants earned income by raising cattle and other animals. Now in the host community, due to reasons such as limited space of the yards, there is no condition for family animal breeding. The lack of production funds has also prevented scale animal breeding from being put into action. Most of the elderly migrants have basically given up the idea of earning income through animal breeding. A villager surnamed M from Qihai Village explained, "Before moving downhill, the grain we grew was enough for us to eat and there was surplus. We raised a few cows and a dozen sheep every year, which brought us ten to twenty thousand yuan. Now we have moved down. The hygienic conditions are better, but the spaces are too small to raise (livestock)."

Some other elderly migrants try to earn income by working as migrant workers. As data indicate, in 2015, 30.8% of the elderly migrants had income from working as migrant workers. But compared with young adults, the elderly ones are disadvantaged in the labor market and are hardly able to find a job since they are physically weaker and are less skillful and competitive. Even though a few elderly migrants are still able to work as migrant workers, this way of earning income is hardly sustainable. As a migrant surnamed L from Hongya Village said, "When we just moved here, I was in good enough health to do some physical labor, picking up odd job on the construction site. These years I've been out of shape and have got sick. I can't do anything and I have little income."

Family-based endowment is still the most acceptable model of endowment to the overwhelming majority of the elderly migrants so far; yet, this model of endowment requires adequate financial affordability and strong filial piety on the part of the children. As discovered in the survey, the problem of elderly migrants not being provided for as a result of either the unaffordability or the lack of filial piety on the part of the children is quite prominent, especially for the incapacitated elderly migrants, who have to rely on social relief allowances to maintain basic livelihood. As a migrant surnamed H from Lihua Village remarked, "We Hui people generally rely on the youngest son to support us when we are old. Now my youngest son and his wife are both out to work. In these years, it has been hard to find a job outside. What they earn is not enough even for themselves and there isn't much surplus to give to us. It's already good enough that they don't ask for money from us."

In terms of expenditure, the daily expenses of the elderly migrants are mainly spent on food, medical care, weddings and funerals and looking after their grandchildren. For the elderly migrants, food takes up a small proportion of their daily expenses. As most of the elderly migrants still engage in agricultural production, the grain they produce is basically enough for the food and drink needs of the family. But for the land-lost elderly migrants, a lot is spent on food. As a migrant surnamed H from Xinghai Village told us, "We had land back in our old home before. Although it was hilly land, the grains we harvest was enough to feed the family and there could be surplus. Since we moved here, we have had to pay for both grains and vegetables. That is very costly."

Medical expenditure is also a standing item in the daily expenses of the elderly migrants. So far, most of the migrants have enrolled in the basic medical insurance for urban and rural residents, from which they can obtain a relatively high proportion of reimbursement if they are hospitalized. But for daily non-hospitalization medical expenses, the elderly migrants still have to pay most or even all the expenses. This is particularly true for those who suffer from chronic diseases and have to take medicine constantly—the monthly medical expenses take a larger portion of their daily expenditure. A migrant surnamed N from Dahe Village said, "Both my wife and I have chronic diseases and we need to take medicine every month. These medicines are not reimbursable, but we can't go without them. We try to spend as little as possible, fearing that we won't have enough money for the medicines."

Elderly migrants attach great importance to major family events such as weddings and funerals. Although these events do not occur very often, once they do occur, it means a reduction in the family finance, sometimes even heavy indebtedness to a family, which is usually borne by the elderly migrants. This will ultimately compromise the endowment quality for the elderly migrants. In Guyuan area, it is a custom for the bride's family to request a large sum of money from the bridegroom's family if they want to marry their daughter. Generally, the bridegroom's family has to pay one or two hundred thousand yuan as the bride price. Although the migrants have moved out of Guyuan, the wedding custom is basically kept unchanged. As a migrant surnamed M from Huaxi Village told us, "Last year my son got married. The girl is from Haiyuan. We paid 150,000 yuan as the bride price. Plus money paid for miscellaneous stuff, we spent more than 200,000 yuan for the marriage alone without taking the house into account. We spent all the money we had and borrowed some 50,000–60,000 yuan from our relatives."

After relocation, most of the young adults work outside all year long. The responsibility of parenting children has totally fallen upon shoulders of the elderly migrants who stay behind. In addition to paying for their own daily expenses, the elderly migrants also have to take care of their children's children. A migrant surnamed M from Mulan village said, "Both my son and his wife are out working. Their children are left for us to look after. It's OK if we are just to look after the children. But the thing is, we have to spend money for the children. Now children are very precious and money is spent on them every day, which amounts to several hundred yuan every month."

As analyzed above, at this stage, the elderly migrants have fallen into serious endowment predicaments. First, due to restrictions of natural resources and the conditions of the migrants themselves, some of the elderly migrants have difficulty in earning income through planting or animal breeding, nor are they able to earn stable labor income by working out. The weakening of traditional "filial piety" has also resulted in the inadequate financial and livelihood support provided by the children to their elderly parents. Second, the loss of land, the incompleteness of medical insurance policies, the continuation of traditional customs and the transfer of family responsibility have led to considerable living expenses for the elderly migrants. With low income and high expenditure, most of the elderly migrants have no choice but rely on the basic pension provided by the government for subsistence and maintain a fairly low living level. To change the endowment predicaments for the elderly migrants, efforts must be made to increase their income. Yet, given the restrictions of natural resources and their own conditions, it is not recommendable to help the elderly migrants earn income by developing planting and animal breeding or labor industries. Rather, the key is to optimize the personal endowment security mechanism for migrants and find out a new road of combining dividends from project revenues, subsidies from the government and payment by the migrants themselves. Only in this way is it possible to resolve the endowment predicaments for the elderly migrants.

6.2 Necessity of Establishing a Personal Endowment Security Mechanism for Ecological Migrants

6.2.1 Establish a Personal Endowment Insurance System for Migrants, Enhance the Elderly Migrants' Confidence in Life

The ecological migration of Ningxia is a large-scale government-led government-sponsored migration programs endeavor. Migration through relocation has led to changes in the social environment and the way of production and life, causing the migrants to lose part of their self-security ability. After relocation, elderly migrants are faced with various unforeseeable economic risks. They have neither adequate economic sources nor the ability to meet their basic living needs. In order to ensure successful implementation of ecological migration, the first thing to do is to find a way out for the endowment problem of migrants, establish a personal endowment insurance mechanism for migrants and enhance the elderly migrants' confidence in life.

6.2.2 Establish an Optimized Migrant Personal Endowment Insurance Model, Ensure that the Migrants Remain Stable

Specialized migration work administration agencies are established from the central to local government levels, which have done a lot of work with respect to the planning of migrant resettlement, the implementation of migrant relocation and the development of follow-up industries after migration. However, due to unfavorable conditions such as the absence and inefficient implementation of migrant resettlement compensation policies, weak economic foundation, inconvenient transport and communication, backward infrastructure and technical management and low personal qualities, the migrants' production and living standard have not been improved or upgraded significantly. Some families are even hardly able to provide for their minimum production and living needs. This is particularly true for some of the elderly migrants. Their living level is even worse than it was before relocation. There are frequent cases where migrants move back to original places or make a second relocation. Hence, it is necessary to establish an optimized migrant personal endowment insurance framework combining one-off payment and continued personal payment so that the elderly migrants can enjoy endowment security immediately after relocation and settle down without worrying about their old-age problems.

6.2.3 Free the Migrants of Their Worries Concerning Resettlement and Follow-Up Development, Enable Them to Devote Their Heart and Soul to Poverty-Eradicating Production

Ecological migration and relocation cover all age groups. For the young- and middle-aged migrants, after relocation, they are not only faced with how to develop production, but also the responsibility of supporting the old and young. As most of the migrants are undereducated with poor capability for self-development; their families are not financially sound and after relocation, they have to put in a lot of money in their children's education, family endowment and medical insurances, their burdens are increased. This has brought difficulties to some migrants in their production and life, severely handicapping their follow-up development. Establishing a personal endowment security mechanism for migrants will help relieve the economic burden of the migrants with regard to family endowment support so that they have more money for economic production. This will become a strong guarantee for the migrants to get rid of poverty and become better off.

6.2.4 Build Migrant Personal Endowment Insurance into a Demonstration Project, Produce Positive Demonstrative Effect

So far, the large mass of ecological migration projects implemented in the central west of China has generated tens of thousands of migrants. They have made contributions to the ecological environment restoration and economic construction of the country, but their endowment security problem has not been well resolved. This is particularly manifested in the low payment standard for endowment insurance, limited raising channels for endowment insurance funds, improper proportions of allocation and lack of risk aversion measures for endowment insurance funds. In view of these shortfalls, an effective option would be to establish a personal endowment security mechanism for migrants that is applauded by the migrants, accepted by the government and affordable by the owner, taking into consideration the basic personal living level of the migrants, the economic development level of the resettlement area, the financial capacity of the government and the income level of the project owner, build this mechanism into a demonstration project for the proper resettlement and follow-up development of the migrants and provide reference for the development of migrants personal endowment insurance in related fields and areas nationwide.

6.3 Principles for Establishing a Personal Endowment Security Mechanism for Ecological Migrants

6.3.1 *Formulate a Personal Endowment Insurance System for Migrants According to Law*

Complying with the relevant laws, regulations and acts is an integral part of establishing an endowment insurance system for migrants. Reservoir migration-related policies currently implemented in China mainly include the "Opinions on the Implementation of the Pilot Project of Relocation for Poverty Alleviation," "Circular of the Ministry of Labor and Social Security of the People's Republic of China," "Regulations on Reforesting Cultivated Land" and "Guiding Opinions of the State Council on the Pilot Project of New Rural Social Endowment Insurance." These policies and regulations are closely associated with the social endowment security system for ecological migrants and therefore should be complied with when designing the endowment insurance system.

6.3.2 *Properly Improve the Migrant Security Level*

The implementation of ecological migration is largely dependent on the sacrifices made by the migrants, and it is just right for these sacrifices to be rewarded. Therefore, when designing the endowment insurance system for migrants, it is important to ensure that the basic living level of the migrants is at least not lower than before relocation. In the meantime, the endowment security level should agree with the economic development level of the resettlement place; the local price and other economic aspects should be substantially considered so that the payment standard for migrants endowment insurance is neither behind the economic development level of the local area nor beyond the affordability of the local government.

6.3.3 *Dynamic Adjustment*

The economic development level of the migrant resettlement area is not static. The principle of ensuring match between the endowment security level of the migrants and the local endowment security level determines that there must be a degree of dynamism in the endowment security level of the migrants. It is necessary to measure and timely adjust the endowment insurance contribution level of the target year with a dynamic model against the payment standard for the migrant endowment insurance for the base year to ensure that the migrants are truly provided for at old age and can share the fruit of economic development in the dynamically changing socioeconomic development.

6.3.4 Proper Share

Individual migrants, central and local governments, collective economic organizations and project owners should share responsibilities properly with matching rights and obligations. A new road should be found that combines subsidies from the government, dividends from project revenues, supplements from collective economic organizations, payment by migrants themselves and their children, social pooling accounts and personal accounts and enhances the rolling development of migrant endowment.

6.3.5 Share of Project Revenues by Migrants

Through relocation, the migrants have made sacrifices for the ecological restoration and regional economic development of Ningxia. They ought to enjoy the benefits from the related project in the original place. Hence, it is appropriate to withdraw a portion of the profits from the related project and use it as the migrant endowment insurance fund.

6.3.6 Compulsory Payment

Due to restrictions of the economic development level, some of the migrants are not so willing to participate in the insurance; the rural collective economic organization and the local government are also reluctant to pay the endowment insurance contributions for the migrants. This is bound to affect the immediate interests of the migrant population. Therefore, toward the basic endowment insurance of the migrants, a compulsory payment principle should be implemented. For the part beyond the basic need, an approach combining government leading and migrant willingness should be implemented to guide the migrants to universally participate in the insurance and improve the security level.

6.3.7 Territorial Management

To avoid the disadvantages of over-lapping management, the endowment insurance funds for ecological migrants should be managed and operated in a centralized manner. The government of Ningxia Hui Autonomous Region should determine the basic principles and main policies; individual local governments should formulate concrete methods to implement territorial management over participated migrants.

6.4 Recommended Innovations on the Endowment Security Mechanism for Ecological Migrants

6.4.1 Object of Migrant Endowment Insurance

The migrant endowment insurance policy should follow the principle of "supporting the old, assisting the middle-aged and promoting the young." That is, basic living security arrangement should be provided for migrants at or above 60 for male or at or above 55 for female; basic endowment insurance should be provided for migrants above 16 but below 60 for male or below 55 for female. Migrants qualified for participating in the basic endowment insurance are separated into three categories: (a) migrants at or above 60 for male or at or above 55 for female who participate in the basic living security arrangement, namely endowment migrants; (b) migrants above 16 but below 60 for male or below 55 for female, namely young- and middle-aged labor migrants; and (c) migrants who are heavily disabled and have lost the ability to work, namely minimum endowment migrants.

6.4.2 Contributions Standard for Personal Basic Endowment Insurance of Migrants

The purpose of endowment insurance is to secure the basic living needs of the elderly and provide a stable, reliable source of income for them. For ecological migrants, the endowment insurance system is a social security system established to secure their basic livelihood when they are old. Therefore, the payment standard for the endowment insurance should meet their basic living needs and be based on the migrants' living expenditure as the basic living standard. After an ecological migration project is implemented, the migrants have lost their original land and other living resources and become a relatively disadvantaged population. They are less able to resist various social risks and also have difficulty in production and life. Therefore, instead of simply using the average living expenditure of the migrants as their basic living standard, their endowment security level should also be improved properly. Hence, it is advisable to set the monthly payment standard for the endowment insurance of the migrants as the per capita monthly living expenditure for the above-average families, namely 60% of the amount of the per capita monthly living expenditure for the migrant families.

The static model used for measuring the personal basic endowment insurance contributions of the migrant is:

$$W = (Y_1 + Y_2 = Y_3 + Y_4 + Y_5 + Y_6 + Y_7 + Y_8) \times 60\%, \quad (6.1)$$

where W is the monthly endowment insurance contributions of the migrant; Y_1 = spending on clothes; Y_2 = spending on dwelling; Y_3 = spending on transport and communication; Y_4 = spending on food; Y_5 = spending on cultural entertainment; Y_6 = spending on household equipment, appliances and services; Y_6 = spending on medical care; Y_8 = other expenditures.

The personal basic endowment insurance contributions of the migrant are not static and should be properly adjusted from time to time according to economic development and price changes to ensure that the migrants can afford their personal basic endowment insurance payment. For this reason, the consumption price index P_i is introduced according to the dynamic adjustment principle.

$$P_i = P_{is}/P_{io}, \quad (6.2)$$

where P_{is} is the consumption price index of the migrant resettlement area for the target year and P_{io} is the consumption price index of the migrant resettlement area for the base year.

By introducing the consumption price index P_{is}, the dynamic model for measuring the personal basic endowment insurance contributions of the migrants as follows:

$$W_{dynamic} = (Y_1 + Y_2 + Y_3 + Y_4 + Y_5 + Y_6 + Y_7 + Y_8) \times P_i \times 60\% \quad (6.3)$$

At the standard sociological sampling ratio of 1–3%, 804 samples are randomly sampled out of migrants at or above 6 of age. Based on the result of survey, the per capita monthly living expenditure of 1179 yuan for 2015 is introduced one-off into Eq. (6.1); we have:

$$W_{2015} = 1179 \times 60\% \approx 707.$$

Therefore, according to the per capital annual spending of the migrants in 2015, the payment standard for the personal basic endowment insurance of the migrants in 2015 was 707 yuan. After that, the payment standard for the endowment insurance of the year should be properly adjusted according to Eq. (6.2) b price changes.

After a migrant has paid the total endowment insurance contributions for the year one-off, which is calculated by multiplying the payment standard for the endowment insurance determined for 2015 with the number of months for which the pension should be received, he will directly enjoy the pension from the government. The number of months for which the pension should be received is the number of months of the remaining lifespan calculated from 60 years of age. In 2015, the average lifespan in Ningxia was 75. Based on this, the number of months for which the pension of the migrant is receivable is 180. The total amount of endowment insurance contributions payable by each migrant is 127,260 yuan.

6.4.3 Sources of Fund for Personal Endowment Insurance of the Migrants

The sources of fund for personal endowment insurance of the migrants include the amount paid by the migrants themselves and their children, project revenues and payment by the state and the government of the migrant resettlement area, reflecting a multipartite sharing principle.

6.4.3.1 Fund of the Revenue from Ecological Compensation Projects in the Original Place Available for Paying Endowment Insurance Contributions

So far, idle land in the original places of the migrants has basically been used for ecological restoration. The autonomous region can implement ecological compensation projects in these places, such as Jinghe River Basin Inter-Provincial Water Environment Ecological Compensation Project and Liupanshan Forest Carbon Sink Market Transaction Project. Through the implementation of ecological compensation projects, ecological compensation funds can be obtained from the surrounding provinces or from other provinces of the country; a portion of money can be used to pay the migrants endowment insurance contributions.

6.4.3.2 Fund of the State and the Government of the Migrant Resettlement Area Available for Paying Endowment Insurance Contributions

The source of fund for the state and the government of the migrant resettlement area to pay endowment insurance contributions mainly comes from the special financial outlay of the government for coordinating urban and rural resident endowment insurance. In 2015, the urban and rural basic endowment insurance in Ningxia was renewed to 134 yuan per person per month. This portion of money can be used for subsidizing the endowment insurance contributions of elderly migrants.

6.4.3.3 Fund of Migrants Themselves Available for Paying Endowment Insurance Contributions

The amount paid by the migrants themselves comes from two sources. One is the compensation for the "four side" woods in the original area of the migrants and the compensation rewarded to relocated farmers based on the policies of returning farmland to forests and returning animal breeding grounds to pastures in their original places of residence provided at the time of relocation in accordance with the "Notice of the People's Government of the Autonomous Region on Printing and Distributing the Opinions on Strengthening Ecological Restoration and Construction in the Areas where ecological migrants move out." The other is the migrants' own money. All money paid will go entirely to the personal account of the migrants.

In respect of inadequate support from the migrants' children, a family endowment filial piety fund can be set up as a supplement to the endowment security mechanism for migrants. Technically, for each migrant above 60, their children should pay no less than 100 yuan per month as an old-age support money for their parents. A special fund should be setup with the money provided by the county-level government of the host area and donated by benevolent companies and individuals to subsidize or reward at a given ratio the children who pay filial piety money for their parents every month.

6.4.4 Willingness of Young- and Middle-Aged Migrants to Continue Payment of Insurance Contributions and the Payment Standard

For young- and middle-aged labor migrants, affected by economic development and price changes, on the one hand, they may become less able to afford the one-off payment of endowment insurance; on the other hand, the endowment insurance contributions paid one-off is hardly enough to meet the migrants' increasing material and cultural needs. Therefore, a survey on the willingness of young- and middle-aged labor migrants to continue payment of insurance contributions will help determine whether a combination of one-off payment and continued personal payment will be adopted and set an appropriate basic payment standard.

6.4.4.1 Survey on the Willingness of Migrants Themselves to Continue Payment

From Table 6.1, among all migrants interviewed, 86.81% clearly confirm that they are willing to continue payment of insurance contributions. This indicates that most of the migrants are willing to further improve their old-aged living quality by continuing to pay endowment insurance contributions.

Table 6.1 Statistics of the willingness of migrants to continue payment (%)

Willingness	Very high	High	Low	Very low	Not sure
Proportion	26.74	60.67	6.96	2.00	4.23

6.4.4.2 Survey on the Contributions Standard for Migrants Themselves to Continue Payment

Currently, according to the "Opinions of the People's Government of Ningxia Autonomous Region on the Overall Planning of Social Endowment Insurance for Urban and Rural Residents" ("Opinions"), the personal payment standard is set with 12 grades spanning from 100 to 2000 yuan. However, this payment standard is based on the minimum basic pension standard of 100 yuan for the autonomous region, whereas the basic pension standard of the migrants is 707 yuan, which is far higher than the minimum pension standard for Ningxia. Therefore, the 12-grade standard may not necessarily agree with the actual needs of the migrants. According to a survey on the migrants wishing to continue payment as to the amount of money they would like to pay per month, nearly 95% say they would like to pay 100 to 1400 yuan per month. Based on this result, the standard for the migrants to continue payment can be designed. Consequently, the standard for the subsidies given by the government of the migrant resettlement place can be set according to the respective payment standards. The exact standard for continued payment ranges from 100 to 1400 yuan, covering 14 grades: 100, 200, 300, 400, 500, 600, 700, 800, 900, 1000, 1100, 1200, 1300 and 1400 yuan. Government subsidies can be rated by reference to the subsidies standard in the "Opinions," that is, at 30 yuan for the 100 yuan grade and 10 yuan for each additional 100 yuan. For grades above 1000 yuan, an additional 5 yuan will be provided for each additional 100 yuan (as detailed in Table 6.2).

As discussed above, for endowment migrants and minimum endowment migrants, one-off payment should be made and included in the pooling account and pensions should be drawn on a monthly basis starting from the second month of payment. For young- and middle-aged migrants, one-off payment and continued personal payment should both be made. The former will be included in the pooling account and the latter in the personal account. Pensions should be drawn from the pooling account and the personal account on a monthly basis after the migrants reach the endowment age.

Table 6.2 Government subsidies standard by payment grade (yuan)

Payment grade	100	200	300	400	500	600	700	800	900	1000	1100	1200	1300	1400
Subsidies standard	30	40	50	60	70	80	90	100	110	120	125	130	135	140

6.4.5 Schemes for Raising Funds for Personal Endowment Insurance Contributions of the Migrants

6.4.5.1 Pooling Account-Based Funding for Endowment Insurance Contributions

The endowment insurance contributions in the pooling account should be funded through a combination of dividends from project revenues, payment by migrants themselves and subsidies from the government. But the amortization ratio should comply with the principle of "less from the damaged party, more from the benefited party." As the migrants are the interest-damaged party, they should share no more than 30%; the government as the benefiting party should share more than 30%. Restricted by the total amount of money affordable by the three parties, the personal payment of the migrants, project revenue and government subsidies should share 20, 30 and 50% of the migrants personal basic endowment insurance contributions.

6.4.5.2 Personal Account-Based Fund Raising for Endowment Insurance Contributions

According to provisions concerning the overall planning of social endowment insurance for urban and rural residents, residents should participate on a self-initiated basis. Personal-account endowment insurance funds should be raised through payment by residents themselves and subsidies from the government provided at different standards. The payment by the migrants should be totally borne by the migrants themselves, at a payment standard determined by the financial affordability of the migrants. The personal payment standard set according to the willingness of the migrants includes 14 grades spanning from 100 to 1400 yuan, for which the government subsidies range from 30 to 140 yuan. As to the pension payment in Ningxia, by reference to the national and local provisions concerning the basic endowment insurance for urban and rural residents, the monthly payment standard for personal-account pension is the total amount of deposit in the personal account divided by 139. After receiving pension for 139 months, pensions will continue to be paid from the government account.

6.4.6 Transfer and Renewal, Operation and Regulation of the Personal Endowment Insurance Fund of the Migrants

Given the particularity of ecological migrants, it is necessary to set a separate migrants personal endowment insurance fund that is separately operated and managed.

6.4.6.1 Transfer and Renewing of the Personal Endowment Insurance Fund of the Migrants

The one-off money paid by the migrants should be included in the pooling account. The money paid one-off is compulsory and benefits each migrant above 16 of age. Therefore, the money paid one-off does not cover the endowment insurance payment made by the migrants before relocation. For those willing to continue payment, the money already paid should be returned to the source account and a new account should be set. That is, the social security department should refund all the endowment insurance contributions paid by the migrants before relocation. The returned money should only include the portion of the money paid by the migrants themselves and the proceeds thereon. Migrants wishing to continue payment may include this returned portion of the endowment insurance contributions in their personal account as part of the continued payment of the personal endowment insurance for resettled migrants.

6.4.6.2 Operation and Management of the Personal Endowment Insurance Fund of the Migrants

An administrative agency specifically in charge of the personal endowment insurance fund of the migrants may be set up by the government of Ningxia Hui Autonomous Region. Such agency will entrust the fund to a commercial bank for management. The commercial bank will select a qualified private institution to make investment with the fund and ensure that the values of the fund are preserved and increased. The ownership and operation of the fund will be separated. The legal person of the fund is the government's social security fund administrative agency; the custodian of the fund is the commercial bank; the investor of the fund is a qualified private investment institution.

6.4.6.3 Regulation of the Personal Endowment Insurance Fund of the Migrants

A separate regulatory system should be established. A regulatory board composed of the government fund administrative agency of Ningxia Hui Autonomous Region, migrant representatives and members of other philanthropic organizations will be organized to oversee the operation, management and allocation of the fund, protect the interests of all interested parties and reinforce the supervision of financial, auditing, judicial and supervisory departments and public opinions.

That all aged people are properly provided for is one of the major goals of building a harmonious migrant new village. At this stage, the ecological migrants in Ningxia are suffering from multiple endowment predicaments. Only by continuously exploring innovations on the endowment security mechanism for migrants is it possible to ensure that the migrants are financially secured, spiritually contented and physically reassured at old age. In the meantime, efforts should also be exploited to inspire the migrants' children to attend to the life of and give spiritual comfort to their old-aged parents and perform their obligation of supporting the old. Only with the joint effort of the whole society in building an endowment security mechanism for migrants is it possible to lift the migrants from fears and worries about their future.

Acknowledgements Data in this chapter come from the "Social Adaption and Follow-up Industry Development of Ecological Migrants in Ningxia" co-sponsored and jointly implemented by the Institute of Sociology of Chinese Academy of Social Sciences and Institute of Sociology and Ethnology of Beijing University of Nationalities. This project covered 10 migrant villages in 3 cities and 5 counties/districts—In Yinchuan City: Mulan and Yuanlong Villages of Yongning County's Minning Town, Tuanjie and Xinhua Villages of Xixia District's Zhenbeibao Town; in Wuzhong City: Dahe and Hongya Villages of Hongsipu District's Dahe Township and Lihua and Chaoyang Villages of Hongsipu Town; in Zhongwei City: Xinghai Village of Shapotou District's Xuanhe Town and Qihai Village of Zhongning County's Dazhanchang Town. A total of 804 questionnaires were recovered.

Chapter 7
Industry Development and Poverty Issues Associated with Ecological Migration

Lili Xun

7.1 Ecological Migration and Poverty Alleviation Through Industrial Development

Since the beginning of the twenty-first century, in pace with the West Development Initiative, the idea of "ecological governance" has received a great deal of attention from policymakers and academics. The emergence of the "green" authority symbolized by the term "ecology" is associated with the following two incidents: The strong sandstorm threatening Beijing and Tianjin in 2000, which turned environmental issues facing the western regions into matters of national security, and the international public opinion of a "green Olympic" when China bid for the Olympic Games in 2001. "Ecological migration" consists of a new objectification means of the state for "border regions", a new "problematization" means, and a new form of state intervention.

"Ecological governance" means that all ecologically vulnerable or environmentally deteriorating areas within the national context are without a doubt the objects of the nation's "topical governance". Broadly speaking, "topical governance" is a package of policy and investment support provided by the central government to local governments for specific "problems". In concrete policy practice, "ecological migration" is the most concerned "ecological governance" option. Migration development is also one of the most important ways of poverty alleviation. Among the "five one-batch" of precision-targeting poverty alleviation, poverty alleviation through relocation is a poverty alleviation measure of the greatest importance. As the western regions are concerned, vulnerable ecology is one of the overriding impoverishing factors in the local area. Ecological migration is not simply a policy approach for ecological governance. It is more essentially an important step toward poverty alleviation development.

L. Xun (✉)
Institute of Sociology, Chinese Academy of Social Sciences, Beijing, China
e-mail: xunll@cass.org.cn

X. Wang (ed.), *Ecological Migration and Targeted Poverty Alleviation in Ningxia*, International Research on Poverty Reduction,
https://doi.org/10.1007/978-981-19-7888-3_7

Ecological migration is in a way a combination of involuntary migration and self-initiated migration. During involuntary migration, there is only "push" but no "pull". Reservoir migration is a typical example of involuntary migration. Those who reside in the zone of dam inundation have no other choice. Hence, these must-do relocations are usually accompanied by greater anxiety and insecurity. In self-initiated migrant relocation, the migrants are usually those who are more readily adaptable to new habitats, such as the young populations. The migration process is also progressive. People can re-establish their lives with their original social and economic ties during this process. But in involuntary migration, people in the affected area are invariably thrown into the same situation at the same time. The risk aversion strategies and social security mechanisms commonly present in the grassroots society will be disrupted. When people are simultaneously thrown into a resource-limited habitat, under great pressure, local counteractions can hardly take effect. Inself-initiated migration caused by development projects are usually the outcome of rational policy programs. These programs use the "greatest good for the largest numbers" as a justified claim. The affected populations are not going to live in the host place for a short while. Instead, they will re-establish their lives on a new land. They have to face a series of problems such as how to build up social relations in the new community and how to establish identify in the new environment.[1,2] The World Bank, Asian Development Bank, and researchers around the world have already reached a consensus of regarding involuntary migration as a profound social concern arising out of certain economic development programs. In the global wave of modernization, induced economy is indeed conducive to improving a nation's productivity. China is a country of migration. According to statistics, over the 50 years following the founding of the People's Republic of China, massive economic constructions had brought about an involuntary migration population of over 40 million. From the 1950s to the 1960s, national interests were paramount and the affected people were required to sacrifice themselves to national projects. The affected people could receive only some symbolic compensation, and it was hard for them to restore and improve their lives after relocation. In the late 1960s, some experts raised an idea of carrying out "migration projects", that is, to boost the development of resettlement areas through migration investments. However, as it was during the Cultural Revolution, the idea of migration projects was not realized in real migration practice. It was not until the 1980s that China came up with a relatively clear conception of resettlement with development. Under this conception, the affected area does not simply mean moving the affected populations out. The affected area should also be included as part of development and redevelopment to allow the migrants to share the good of development. The prerequisite of migration with development is the market principle. "Modernization development" becomes the justified claim for a migration project. Hence, follow-up industries of ecological migration have also become the key to the success of a migration project.

---

[1] Cernea and Guggenheim 1993).

[2] .

Ecological migration bears the characteristics of both involuntary migration and self-initiated migration. On the one hand, ecological migration is an important means of poverty alleviation development, and migration is linked to development. Ecological migration is a paradigm of government-led development. An ecological migration program is usually a development program of a region, involving the modern transitions of urbanization and industrialization. These government-driven development programs often bear a top-down coerciveness. On the other hand, from the subjective standpoint of the settlers, as the original area of ecological migration is basically ecologically vulnerable, resources-limited, economically backward areas, the development prospects of the migration program also carry the development prospects of the locals from economically backward areas. They wish to take part in the modernization development under the promotion of the government and seek new opportunities for family subsistence out of poverty in the wave of urbanization and industrialization. Such a combination of involuntary migration and self-initiated migration has posed quite many uncertainties to the policy practice and anti-poverty effectiveness of ecological migration. The tension between policy planning and policy process and between system design and livelihood practices can both give rise to new poverty problems.

"Ecological migration" is an anti-poverty policy and a development intervention method widely promoted in the western regions. In 2001, the central government set up four "alleviation poverty through relocation" pilot regions. Ningxia is among them. This autonomous region has long had a history of migration. Since the beginning of the 12th FYP period, the ecological migration in Ningxia has displayed some new historical characteristics. The profound integration of ecological migration planning with urbanization and industrialization, in particular, has assigned new characteristics to the impoverishment associated with ecological migration.

7.2 Development Models of Follow-Up Industries and Impoverishment Risks Associated with Ecological Migration

Industry planning and development of ecological migration are the product of government leading. Understanding the roles and functions of the multiple actors involved in the formation of industries is the objective of this section. We have discovered that the industry planning in the ecological migration policy of Ningxia contains two industry forms: Industrial concentration areas and agricultural concentration areas. The agricultural concentration areas are usually transitional areas from traditional agriculture to modern agriculture.

7.2.1 Industrial Development, Labor Migration, and Impoverishment Risks

In the 12th FYP ecological migration of Ningxia, the concentration of migrants, especially labor migrants, toward industrial concentration areas represents a striking feature. In the strong traditional industrial areas in the northern river regions of Yinchuan, such as Jinfeng District and Shizuishan, ecological migration has already become a form of local labor input. Heavy government investments on infrastructure have also attracted numerous young migrants and their core families. In the south, industrial development has been greatly restricted by water resource conditions. But under the promotion of the local government, some low-energy industrial enterprises have also sprouted one after another.

7.2.1.1 Metropolitan Industry Park

In the 12th FYP ecological migration program of Ningxia, labor migration accounts for 25%. The targets of labor migration are young core families with work advantages from the southern mountainous areas of Ningxia. The core purpose is "urbanization". The government pays to build labor resettlement areas in cities along the Yellow River. Labor migrant apartments are built at 40 m^2 per household. After giving up their contracted land in the original place, the labor migrants can obtain an urban hukou as well as access to the appropriate education, medical, and other social security grants. The labor migrants can rent a labor migrant apartment at a low price as a turnover dwelling during the first five years after moving in. Yet, the policy practice of labor migration has been greeted with a range of obstruction. Migrants commonly prefer "ecological migration" rather than "labor migration", because ecological migration provides them with a given amount of land in the host place through land replacement. Although they are settled on agricultural hukou, with land guarantee and the ownership of migrant resettling housing, the farmers are more worry free.

Yinchuan is the center of the along-the-Yellow-River economic area, as well as one of the main inflow destinations for the surplus labor powers from central south Ningxia. Nevertheless, during the 12th FYP period, the city's labor migration program city has almost become abortive. Jinfeng District Industrial Concentration Area (hereinafter, concentration area) was originally one of the resettlement destinations for labor migrants. The concentration area lies in the southwest of Jinfeng District. As planned, the area will contain one zone and two parks. The zone is the "Minor Business Industry Park of Jinfeng District Industrial Concentration Area". This zone extends to Changcheng Mid Road in the north, Liupanshan Mid Road in the south, Tongda South Street in the east, and Baotou-Lanzhou Railway in the west. The planned construction area is 11,115 mu. The two parks are Yinchuan Technology Park and Jinfeng District Comprehensive Industry Park. The former extends to Liupanchan Mid Road in the north, South Ring Expressway in the south, Tongdan

South Street in the east, and Baotou-Lanzhou Railway in the west. The planned construction area is 5278 mu. The latter extends to Yinchuan Funeral Parlor in the north, Lucaowa Detention Basin in the south, Han Cemetery in the west, and Yinzhi Road in the east. The planned construction area is 1154 mu.

Since its kickoff in 2004, the concentration area has always aimed to build a modern hi-tech demonstration area with first-class environment, conglomerated industry strengths, well-organized functionality, and remarkable scale effect. So far, an aggregate of 5 billion yuan has been used for construction, having built a 4-vertical 6-cross road network and associated drainage infrastructure. The park area has been basically equipped with all necessary utilities and transport. A farmer resettlement area, a migrant worker apartment building complex, a public service center, and an entrepreneurship investment building have been constructed. In total, 153 workshop buildings have been constructed in the minor business industry park, totaling 180,000 m^2. In the concentration area, the supporting services are being optimized, the scales expand continuously, the industry aggregation degree keeps rising and the industry carrying capacity continues to increase.

In the first half of 2015, the concentration area reported a gross industrial output of 1.59 billion yuan, up 11.9%, a pretax profit of 143 million yuan, up 10.3%, and a fixed asset investment of 320 million yuan, up 14.3%, from the same period of the previous year. The 15 4S stores hosted in Jinfeng Motor City totaled an operating revenue of 860 million yuan. As of the date of our survey, the park area was housing 227 enterprises, of which 166 (including 85 in the entrepreneurship park) had started production. Scale effect is taking place among the enterprises there. Six industry clusters have already taken shape there. These include an equipment manufacture industry cluster represented by Ningxia Machinery Research Institute, a textile industry cluster represented by Dehong (Ningxia) Textile Co., Ltd., a halal food industry cluster represented by Ningxia Tianrenhe Muslim Beanproducts Co., Ltd, a new building doors, windows and furniture industry cluster represented by Cadillac and Infiniti, and a commercial concrete industry cluster represented by Ningye Tongli Concrete Co., Ltd.

Since 2015, along with the increasing downward pressure on the economy, the enterprises in the industrial park have begun to suspend production. Their labor absorption capacity has declined. The labor migrants are no longer able to work for the enterprises in the park. Some have left again to find other job opportunities. The significance of labor migrant apartments to labor migration has been extremely limited. To some extent, labor migrant apartments can be seen as a public service supply provided by the government for promoting migrants' integration into the city. But the uncertainty of the labor demand market can hardly match perfectly with the certainty of labor migration policy planning. This compromises the eagerness of migrants from the original place to participate and increases the impoverishment risks to the labor migrants.

7.2.1.2 Eastern Industry Transfer

"Longhu Model" is a typical example of industry-guided labor migration in Ningxia. Longhu Economic Zone used to be a labor migration diaozhuang (a place receiving whole-village labor migrants) for Longde County identified by the party committee and government of the autonomous region in 1983 according to the poverty alleviation development guideline of "assisting the mountainous area with resources of river areas, inter-assistance between mountainous and river areas". In 1992, it became an autonomous region-level poverty alleviation economic development zone with the approval of the people's government of the autonomous region. In 2003 and 2004, it was placed under the administration of Shizuishan City and Dawukou District, respectively. In 2009, it was established as Xinghai Town. The development zone is situated on the southwest side of Dawukou District, facing the Baotou-Lanzhou Railway to the east and bordering Shahu Tourist Scenic Area to the south and Xinghai Lake to the northwest. It has superior geographical advantages and convenient transportation. It has a total area of 140 km^2, including 12,700 mu of farmland, and its administration encompasses 8 administrative villages and 3 communities with a population of 60,000. Within the zone, there are 17 above-scale enterprises, and 6 schools and some 8,000 school students. Business tax administrations, public security, hospital, school, court, post office, agricultural bank, and credit cooperative all have their representations in the development zone. In 2014, the disposable income of the rural residents was 8560 yuan, up 9.5% from the previous year. The development zone has been honored the "5th National Modal Collective for Ethnic Unity and Progress", "National Eco-civilization Demonstration Town", "Home of Chinese Folk Culture and Art", and "Advanced Unit of Urban-Rural Social Endowment Insurance Work of the Autonomous Region".

Recently, the development zone has invested great efforts in advancing the coordinated urban–rural development strategy. Sticking to the decision and arrangement of the municipal party committee and municipal government of building "three economic zones", they have planned and constructed Xinghai Economic Zone with high standard, high start point, and urbanization. First, they coordinated the urban–rural planning. According to the development positioning of "cultural window, water town north of the Great Wall", they spent 2 million yuan and completed the master planning and regulatory detailed plan for Xinghai Economic Zone. Its toponym planning work takes the lead in the city and even the autonomous region. Second, they coordinated the urban–rural infrastructure constructions. According to the conception of "centralize large settlements, build large communities", they implemented Xinghai Jiayuan, Kangju Jiayuan, and Longxianyuan projects. They completed the reconstruction of 1500 old dwellings and the construction of a 10,000-m^2 cultural and sports plaza. They implemented the first rural people drinking project in the autonomous region. The production and living conditions of residents in the jurisdiction have been greatly improved.

They planned and constructed rural infrastructure to the urban standards, having constructed ten key urban roads, including Shahu Avenue, Xingguang Avenue, Longhu Avenue, and Yushantan Street, totaling 120 km with an investment of almost

1 billion yuan. They coordinated the opening of bus lines from downtown Dawukou to Xinghai Economic Zone. A crisscrossed, all-directional transport framework has taken form in the area. So far, water supply, telecommunication, digital television, and Internet have connected to all parts of the area. One-station/six-station rainwater collection projects, natural gas regulation station and LPG storage and distribution station relocation projects have all been completed. The overall service functionality of municipal facilities has been fundamentally upgraded.

In 2011, Xinghai Town became Dawukou District labor resettlement destination. "By the lake, close to roads and near the city" characterizes the Mu'en Xinju neighborhood in Dawukou resettlement area. Here, 2514 suites of turnover housing totaling 130,400 m^2, floor area have been built. Unlike the labor resettlement housing allocation policy of Yinchuan, the migrant families in Mu'en Xinju from Longde County had the option to own their respective turnover dwellings in five years at the price of 12,800 yuan each. All labor migrants from Longde have given up their land and dwellings back in their home villages. Relocation of the migrants in Xinghai Town greatly has added to its labor advantage. The textile factory, which used to be in Shandong, has had its new factory built near the labor resettlement area. The employment rate of the women of the migrant families has been greatly increased. Dawukou has a high inflow rate of labor migrants. But in terms of integrating into the city, the labor migrants have not fully become a citizen. As the Shizuishan government is financially limited, public service for the newcomer populations cannot catch up in a short time. So far, the migrants are still able to obtain an urban hukou. Their children's education and medical security cannot be realized. This has greatly embarrassed troubled the life choice of the new migrants. As the policy design and interconnection is incomplete, the migrants living costs and livelihood risks have been increased. Families with three generations having to reside together have also left Dawukou to seek a livelihood in other places again.

7.2.1.3 Industrial Park in the Mountains

Intra-county labor migration in the mountains of south Ningxia is also an important form of labor migration and a new type of urbanization. Mining Industrial Park of Xiji County is very representative of the industrial development and labor migration in the southern mountainous areas. Xiji has always been a labor output county. Its labor export population is over 120,000, making the so-called rural migrant economy. Here, labor income is regarded as the "iron crop" for farmers' income. Restricted by limited water resources and other conditions, industry has long been a blank in this county. Moreover, although agricultural resources here are relatively abundant—for instance, Xiji is known as China's home to potatoes, deep processing of farm produce has always lagged behind. Industrial resources are deficient. Farm produce can hardly be converted with high added value. This has become a bottleneck restricting the economic development of this part of the autonomous region. Taking advantage of the "inter-assistance between east and west" policy, Mining Industrial Park has established an industrial development model with local characteristics by virtue of investments from Fujian manufacturers.

Minning Industrial Park lies in Shuiquan and Yuanhe Villages of Jiqiang Town, the county seat of Jixi, with a planning land area of 12 km^2. Built according to the overall space functional disposition of "one park, five zones": a distinctive farm produce processing zone, a light industry zone, an electronics industry zone, a logistic zone, and a business service zone, the industry emphatically develops fine and deep processing of high-quality potatoes, carrots, tomatoes, and halal meat products, the manufacturing of agricultural materials, folk costumes and electronic products, together with business service and logistics providing services for the enterprises inside the park. The Minsheng Jiayuan neighborhood next to Mining Industry Park is a low-rent housing project built for the labor migrants in this park during the 12th FYP period, where a total of 1224 suites of homes are planned. In the eyes of the manufacturers, Minsheng Jiayuan is undoubtedly their "employee dormitories".

The industrial park in the mountains is close to both the raw stuff origin and labor-intensive area. Its development is theoretically highly beneficial to the poor. In the survey, we discovered that this process is not as easy as we imagined. We even discovered that not all the manufacturers in the industrial park are totally willing to consider local labor first. Behind this phenomenon, cultural conflict is a principal factor. On the one hand, the manufacturers would like to pay on a piecework basis according to the market rule, but the local workers are more willing to accept a minimum wage payment. For enterprises, they are the market player and their order-based production requires a flexible labor employment. For the workers, being flexible will mean employment risks and uncertainties. On the other hand, the mountains in south Ningxia are a Hui-inhabited area. The features of ethnic culture have also brought the enterprises quite a lot of hard nuts to crack when dealing with labor-capital relations. Many manufacturers complain that the local workers are opposed to working extra hours. The complicated informal relations also cause the workers to refuse to obey the authority of their boss, leading to disorders in production management.

Summing up the above, in our survey, we discovered that the transition from farmers to industry workers usually does not happen in one action. In the industrialization of Hui-inhabited areas, attention should also be paid to the potential influences of ethnic cultural divergences. The east-to-west industrial transfer is grounded on the rich local labor resources in the local area. But one reality is usually overlooked: The local labor powers are often more inserted in a localized community texture and cultural network. For a long time to come, they will remain "semi-transitioned" from farmer to worker. The turnover housing constructed by the government for the intra-county labor migrants is helpful to stabilizing employment and reduce labor costs for enterprises. In future development, more importance should be attached to protecting the entitlements of migrant workers and coordinating the labor-capital relations.

7.2.2 *Modern Agricultural Areas*

Ecological migration destinations are developed into modern agricultural concentration areas: On the one hand, the government-driven land consolidation project has planned out large tracts of land. This is good for the entry and development of agricultural enterprises. On the other hand, inflows of migrants also provide adequate labor for these modern agricultural areas. By taking the geographical advantages of being close to a metropolis, the ecological migrants on the periphery of Yinchuan have developed modern agriculture and formed a distinctive agricultural belt on the periphery of the metropolis.

7.2.2.1 Blurred Subjectivity of the Migrants

The Binhe Jiayuan neighborhood in Yueyahu Township, Xingqing District of Yinchuan is the largest resettlement destination in this "ecological migration". From 2011 to 2015, it had received 4000 households totaling 16,800 residents. These migrants all come from Pengyang County in the southern mountainous areas. The migrants were relocated on a whole-village or whole-group basis. As the local farmers described it, "All those in the marked circle were to leave. It would be meaningless to stay here if everyone else has left". Additionally, the local government also adopted an "uprooting" method to prevent the migrants from coming back—They demolished the migrants' dwellings in the mountains. No sooner had the migrants left than their original dwellings and yards were bulldozed away. Many families who had just built their dwellings two years before had to sell off the building material of their dwellings at a low price.

These "uprooted" farmers have no way back. They, who had used to live separately in the mountains, were centrally settled in the new resettlement area. Each of the migrant families is provided a 54-m^2 dwelling. For this, they pay 12,800 yuan out of the total cost of 37,800 yuan and obtain the ownership of the house in five years. The entire Binhe Jiayuan is squarely, neatly outlaid and divided into four "villages". Each dwelling is marked with the house number. Each new village is usually composed of a number of natural villages or groups back in the mountains. Most of the original village officials have lost their positions. The village director of the new village is still held by a local resident. The party secretary of the new village is a young college-graduate official appointed by the township government.

This uncultivated sandy land on the eastern fringe of Yinchuan City has been revitalized into a new "urban space" by the inflows of the migrants. This space planning in line with the logic of the state "managerialism" undoubtedly well complies with the governor's requirement for a "clear" and "simple" context. Yet the spatial dimension (regionalism) of the multivariant, complicated social practice inside this new community has been heavily overlooked. Nearly a third of the families in the new migration area have three generations residing together. Back in the countryside, the hukou of the older family members was usually together with that of one of the sons. As the resettlement housing is allocated according to hukou and the government has set the timeline for household splitting, for many families that failed to complete household splitting before this timeline, the older members have no separate homes and have to live with their children. As they are already used to residing separately back in the countryside, residing together has become very inconvenient for the family members. To make things worse, as the older members no longer work on land and have to rely on their children for a living, this poses tremendous pressure to their children. Inter-generation conflict is becoming more evident. Additionally, in a new resettlement area, problems not only arise out of the mixed residing between the Hui and Han migrants, but also the centralized residing among different sects of the Islamic religion. In the southern mountainous areas, individual religious sects of the Hui people each have a strong regional character. Different sects have their own mosques in place. Worshiping in the same mosque would be unimaginable to them. Yet in the migration program, theoretically one single mosque is allowed to be built per resettlement. Its siting is also quite close to the Han-inhabited area. For this reason, the siting and number of mosques has become a topic of controversy among the local Hui migrants.

In reality, of the funds invested toward "ecological migration" projects in Ningxia, only a small proportion has been used for resettlement housing. The vast majority of the money has gone toward land consolidation and transformation related to industrial development, irrigation constructions, and constructions of associated infrastructure.

In the ecological migration program, each of the migrant families in Yueyahu is entitled to a cow, and each person one mu of land. The cows are purchased by the government with government funds. As to land, the government spends money to consolidate the land and construct irrigation facilities, turning the state-owned uncultivated sandy land into usable farmland. In exchange, the migrants' land in the mountains will be totally transferred as state-owned land. Unlike the "cow villages" in Inner Mongolia, the migrant families in Yueyahu do not directly participate in farming cows or cultivating land. Instead, the government centralizes these production materials into a Ten Thousand mu Cow Ecological Farming Demonstration Zone in the form of cow trusteeship and land circulation, which is operated by large cow farming enterprises and agricultural development companies. Each migrant family can receive a regular cow income of 2800 yuan per year. Six years later, the farming company will return a high-quality, healthy cow of the same months, same weight, and same birth as the one assigned to them in the beginning of the trusteeship. Besides, each migrant family can also receive a regular land income based on the number of their family members. The fixed income per mu of land is equivalent to the then current price of 150 kg of wheat. For instance, if the family has six members, they possess 6 mu of land. At the price of two yuan per kilo of wheat, they can obtain a regular income of 1800 yuan per year. The Ten Thousand mu Cow Ecological Animal Farming Demonstration Area is an associated industry planning area for the ecological migration destination. The demonstration zone is located 5 km away from Binhe Jiayuan. It covers a total land area of 12,800 mu. Inside the zone, there are 15 standard scale cow farms, one large feedstuff delivery center, one cow technical service center, and one 10,000-mu feedstuff cultivation base. Water and electricity facilities, road, and stables inside the zone, together with the land modification and consolidation of the feedstuff cultivation base, are all constructed with the investment for the ecological migration project. The local government has entered into cooperation with a number of leading companies and introduced large cow farming enterprises into the demonstration zone, with a plan of keeping 36,000 cows on hand. The thousand mu feedstuff cultivation zone leases or circulates the land to agricultural development companies. From another angle, it is because the cow farming demonstration zone is listed as an associated industry for the 4000 ecological migration households in Yueyahu that the government has been able to carry out infrastructure constructions by virtue of the "special funds for ecological migration projects", and to attract leading farming enterprises and cultivation companies into the zone. It isn't hard to understand that, in a sense, the farming enterprises providing each migrant family with an income of 2800 and the agricultural development companies providing 300 yuan for each migrant is a profit sharing with the migrants. From the standpoint of the migrants, cow farming is not a labor-intensive industry and little would be expected if cows were separately farmed by individual households. As to land, given the land conditions in the host area of Yueyahu, agricultural cultivation by individual households would not be sufficient to satisfy food and clothing, either. The government's scale industry integration with the help of leading enterprises can yet be regarded as an effective method for reducing market risks. As the migrants themselves describe it, they are owners of a cow but never see the cow; they are owners of land but never see the land. The government's industry planning has turned the migrants to "idling labor powers". The fixed incomes from cow farming and land circulation are obviously inadequate to meet the normal expenses of a family. The migrants have to work elsewhere to maintain family existence. To some extent, the ecological migrant villages in Binhe Jiayuan have already become an assembly of potential migrant workers. The establishment of ecological migrant villages has produced a labor aggregation effect. Yet increasing numbers of labor powers have led to falling wage price.

The industry planning performed in migration planning did not give adequate play to the subjectivity of the migrants in self-development. Instead, "passive centralization" of land has prevented them from realizing true development participation. Since 2015, as the farming and husbandry enterprises that have contracted the land become unable to pay the migrants their land rents due to poor profitability, the livelihoods of the migrants have been threatened. Following the economic downturn, large numbers of the people in the resettlement area have moved out. Large numbers of houses have been leased to others. The security conditions have begun to worsen.

The "labor advantage" formed in the migrant-inhabited area has become an important chip for the local government in their further investment invitation into the migration project area. Many labor-intensive enterprises are particularly interested in the cheap labor resources there. Next to the Binhe Jiayuan residential neighborhood, the government of Xingqing District has installed 100 solar greenhouses and 300 mobile arch sheds for developing distinctive crop farming. Bridged by Xingqing District Bureau of Labor Employment Services, Ningxia Zhoujing Shirong Flowers and Plants Import Co., Ltd has entered into cooperation with Yueyahu Ecological Migrant

Village to grow carnation in the greenhouse facilities at the settlements. The flower company contracts the greenhouses to the migrant families. The seeds, fertilizers, and pesticides needed for flower growing are all provided by the company. Technical instructions are also provided by technicians from the company. The farmers become contractors of flower growing after paying a sum of money as deposit and attend the flowers every day at the instruction of the technicians. During the harvest season, the company purchases the flowers according to their quality ratings. Contractors with more labor input and higher technical competence will obtain higher incomes. The introduction of Ningxia Zhongyin Cashmere Knitting Factory also follows this suite. The government provides free workshops to the factory. The large number of women in the migrant villages not gainfully employed is the key appeal for the factory. Before starting production, the factory recruited over 400 women workers for training. In order to stabilize the number of workers available for the company, the government paid each of the trainee 22 yuan per day as a subsidy. Many young Hui women go to work at the textile factory.

7.2.2.2 Predicament of Cooperative Development

Heshun Xincun in Xingjing Town of Yinchuan's Jinfeng District is a 12th FYP ecological migration destination. After years of development, it is running into new problems. On the one hand, as the resettlement project was planned in a hurry, the land ownership of the village is not clearly defined. The migrants have not obtained the ownership of their houses there. The house foundation in the resettlement area is shallow and the migrants cannot increase their dwelling space by adding a second story. Many families have put up a separate dwelling next to the vegetable greenhouses to avoid having multiple generations residing in the same house. On the other hand, vegetable growing there has been advantaged by the migrants having to face purchasers separately since the beginning. The vegetable cooperative has not played its role. As the resettlement area is actually a "fill-in" mix of migrants randomly allotted from multiple townships and villages, it is hard for the original elites of individual places to play their roles or for a combined force to be formed within the village. Separate management of greenhouses is gradually replaced by joint management. The migrants have many opportunities to work on the outskirts of cities. In the situation of price instability, many families would prefer to sublet the greenhouses to capable growers and able men instead of investing more on infrastructure. The village has begun to differentiate inside. The village governance of public affairs has also fallen into a predicament. The under-provision of public service by the government in post-migration development, especially the severe lag in social construction-based poverty alleviation actions, has caused the resettlement area to be "de-powered" instead of "empowered". The impoverishment risks have been intensified. Some migrants have begun to return to their home villages to find other ways out for development. Some migrants who are unable to achieve self-development have become new impoverished ones.

In the 1990s, Michael M. Cernea introduced an Impoverishment Risks and Reconstruction model (the IRR model). This model places risk at the core of attention and greatly improves the ability to predict, diagnose, and mitigate impoverishment issues associated with resettlement by identifying the aspects to be addressed throughout the resettlement process through confirmation of risk variables associated with resettlement process. According to this IRR model, there are eight risk variables of impoverishment associated with non-voluntary migration: landlessness, joblessness, homelessness, marginalization, increased morbidity, food insecurity, loss of access to common property, and social disarticulation. Cernea's risk model of non-voluntary migration constitutes an effective tool for analyzing and interpreting the process of non-voluntary migration and provides us with a better insight on the challenges facing resettlement. It is also a valuable guide for planning and policy-making for resettlement-related development projects. As a contribution of sociological and anthropological knowledge to non-voluntary migration and even the entire induced development policy, it has already become a framework for other interested researchers in their studies.

Impoverished migrant families have the following characteristics. First, their human capital is on the low side, namely, the entire family is undereducated. Second, their income is not well structured. Families relying more on agricultural incomes are more prone to impoverishment than those relying on salary incomes or operating incomes. Families relying on transfer payment from migration projects are even more vulnerable than those relying on agricultural incomes. The possibility of families relying on agricultural incomes to become impoverished is 3.6 times that of families relying on salary incomes and 1.73 times that of families relying on operating incomes. Before relocation, the migrants had high expectations of their post-migration lives. Many assumed that migration would raise the level of their lives and took it for granted that the government should be responsible for their post-migration lives. This high dependence on the government is successive and can grow into the "wait, depend, want" concept already familiar to us. This is even more prominent among the impoverished migrant populations. This high dependence mindset can easily give rise to a huge mental gap in the migrants when they are faced with post-migration production or livelihood risks. It also greatly weakens their psychological preparedness to counteract risks in a positive manner.

7.3 Anti-Poverty Intervention and Migrant Settlement Development

When "anti-poverty intervention" becomes a national strategy, the significance of poverty alleviation is already beyond the realm of social security and social welfare. Not only does it stimulate the "cross-disciplinary collaboration" among multiple actors—government, market, and society, it also triggers a recombination among capital, technology, land, and labor, thereby reconstructing the prospects of the local society. The significance of poverty is also endowed with more profound significance of social governance beyond poverty reduction itself. In the practice where "poverty alleviation" serves as a public policy, there always exists an inseparable but expansible topic: The tension between the poor as a "social category" in policy discourse practice and the poor as "actors" in policy intervention processes is always there.

The poor as a "social category" are the product of the poverty alleviation identification and targeting mechanism. For a long time, the income-based poverty identification framework has governed the world's designing and implementation of anti-poverty strategies. "Poverty line" is as an important operating tool in both the measurement of the size of the impoverished populations and the investment of poverty reduction resources. It is fair to say that the poor as a "social category" are usually presented with abstract statistical data.

When we come into a concrete anti-poverty intervention process and face the concrete poor, our understanding can no longer stop at the inadequacy of income. Our understanding of poverty has changed from a singular economic dimension to a comprehensive, multidimensional outlook on poverty. During this process, the contribution of Amartya Sen is of monumental significance. In his view, poverty is not ultimately a matter of income; but a failure to achieve certain minimum capabilities in connection with deprivation and marginalization. These capabilities are presented as entitlements in the society, and this entitlement in turn is determined by social features such as law, economy, and politics. The "poor" cannot be simply regarded as a member of the enormous number of poor people, but a member of a special class, a specific occupation having different resource endowments and dominated by different entitlements. Incomes are important, yet increasing incomes alone do not make an effective action against impoverishment.

Ecological migration is a term unfamiliar to the general public until after 2000. Recently, it has even become a widely popular social policy in the ecologically vulnerable regions in West China. The ecological migration projects in Inner Mongolia Autonomous Region, Qinghai's Sanjiangyuan region, and Ningxia Hui Autonomous Region are particularly notable. The connotation of ecological migration used to be a widely debated topic among researchers. A more cited definition is: "Ecological migration is a migration activity that takes place in response to deteriorating ecological environment or for improving and protecting the ecological environment, and the displaced populations resulted from this activity".[3] While there do exist populations who volunteer to relocate themselves on account of deteriorating natural environment, and ecological migrations widely practiced in the ecologically vulnerable regions in West China are mostly crowned with the name "ecological protection", those implemented over the past few years are more connected with local urbanization and poverty alleviation projects. Many authors also consider ecological migration to be an important "anti-poverty" strategy for ecologically vulnerable, resources-limited areas[4]. Guided by this ideology, huge sums of state funds for ecological governance and poverty alleviation have been used for the infrastructural

[3] Bao (2006).

[4] Ni (2007).

construction in the "whole-village advancement" ecological migration projects. The numbers of the relocated populations are also a magnitude of a million. Worldwide, migration policies are more centered around land acquisition and development rather than income restoration. More importantly, many decision makers or planners have not realized that economic compensation is a measure with great limitations. The compensations provided are usually less than enough for the affected populations to buy back their previous land or other living facilities. The amount of compensation is often underestimated for one reason or another. The laws of many countries do not ensure that people are compensated on a "land for land" basis.

The success of government-led programs has always been regarded as the most important lesson from China's anti-poverty intervention efforts. The top–bottom government administrative system constitutes the greatest and most motivating actor in China's poverty alleviation process. It is fair to say that China's anti-poverty intervention has been inserted into the central-to-local hierarchical and segmental relations since the very beginning. The operating mechanism, organizational cost, and communication efficiency of the bureaucratic system inside the government, as the main player of poverty alleviation transfer, are all direct determinants for how well the objectives of the poverty alleviation policy are fulfilled.

Although many of the poverty alleviation development projects are aimed at "countering poverty", the projects themselves are a process of resources reorganization and social reconstruction. These short-lived political-economic reorganizations usually place a group of people at a marginalized disadvantage, leaving them in constant impoverishment. Hence, apart from bringing "benefits", development can also create "impoverishment". Hence, whose "development" is the "development" becomes a question worth of reflection.[5] In our opinion, how to resolve the social disruption aspect of development is an issue to be intensively studied and prudently decided when designing any ecological migration project. Besides, giving far more emphasis on projects rather than migrants and on relocation rather than resettlement is also a serious problem among ecological migration projects. These policies do not place the income restoration of migrants in a substantially momentous position in their strategic thinking and underestimate the potential economic setbacks that social disruptions can incur to the migrants in their planning schemes. This results in austere difficulties and immeasurable losses to both the implementation of the policies themselves and the subsequent development. All countries try to solve the compensation to migrants through financial means. Yet the financial resources available for the migrants are usually very limited with great financial gap, coupled with the lack of political commitment of the government toward income restoration and the possibly tight budgets and rigid distribution procedures, the financial resources available for migrant reconstruction are extremely limited. So far, all commonly used economic analyses on migration costs tend to internalize the costs instead of externalizing them. That is, reconstruction costs are not included as part of the project budgets. Of course, when reconstruction costs are grossly underestimated, these losses will

[5] Xu and Wang (2000).

be borne by no one else but the displaced migrants themselves. Their falling living level is therefore an inevitable outcome.

After deprived of the basic guarantee for agricultural income, the farmers' ability to resist living risks has vastly reduced. Many families have become impoverished as a result of the high educational expenses for their children and the high medical costs for their diseases, coupled with the instability of nonfarm employment. This will compromise the social justice and equity and the social stability built on this basis.

Migration agencies are important stakeholders in the relocation of migrants. Their roles and abilities play a critical part in the migrant reconstruction work. Michael M. Cernea suggests that migration agencies usually lack policy execution ability, relevant organization ability, and appropriate social working skill. This indirectly adds to the impoverishment risks to the migrants. In China, the institutional framework of migration is a three-tier structure: The first tier comprises the administrative departments of the central government, namely, the State Council and its subordinate ministries/party committees and other working bodies. The second tier comprises the administrative departments in charge of the industry of the provincial, municipal, and county-level governments and relevant government agencies, including the administrative departments in charge of the industry, land management agencies, urban housing demolition and administrative agencies, surveillance and auditing departments, and judicial authorities at all levels. The third tier comprises the owner of the development project (including the surveying and design institutes, supervision, monitoring and assessment institutes and consulting institutes providing services for the owner), and the agency employed by the owner to implement resettlement. These migration agencies are usually "advantaged" relative to the displaced populations. It isn't hard to see that in China, the developers have strong overlap with the institutions in charge of migration. Their are not responsible for making sure the resettlement has sustainable outcomes. They organize the relocation itself, but do not have what is needed to be deeply involved in post-relocation work. Besides, migration agencies also lack effective policies and actions with regard to the social survey and poverty monitoring of the migrants. They usually conduct a pre-migration social survey to affect the release of resettlement funds, but scarcely perform long-term monitoring during post-relocation livelihood restoration and social reconstruction of the migrants. At the beginning of the resettlement, the migrant families are benefiting from rising incomes and improving livelihoods. The migration agencies at this time will tend to assume that the migration project is successful. But as the risk tolerance of the migrants weakens, long-term monitoring would reveal that a considerable proportion of the migrants are already impoverished and have accumulated a lot of discontents. Unfortunately, this long-term, live poverty monitoring mechanism is extremely scarce. Even if a social survey is arranged before migration, due to lack of the involvement of sociologists and anthropologists, excessively more importance is attached to measuring the family income conditions to facilitate monetary compensation than the social organization aspects of families and communities. This leaves a large dead zone for the design of migration programs. To make things worse, a few migration agencies even use their power to manipulate behind the scene and misappropriate the resettlement funds. This not only further reduces the already limited amount of resettlement compensation, but will also arouse strong discontents and objections from the migrants, consequently putting social stability at risk.[6] In China, the principal migration agencies are the administrative agencies at different levels. Affected by a long-standing planning system, these migration agencies are accustomed to a top-down form of action. The opinions and ideas of the affected populations can hardly find their way into the decision making and execution of policies. Nevertheless, this top-down form of action within administrative agencies usually overlooks the diverse characteristics and demands of the grassroots communities. A planning program created without substantial knowledge of local information will be hardly able to adapt to the local conditions. Its subsequent failure will be but an inescapable outcome. Moreover, the authoritarianism of migration agencies also determines the lack of channels for the migrants to voice their opinions and complaints. The problems encountered by the migrants when trying to re-establish their lives cannot be addressed and resolved in a timely manner. Eventually, the long-accumulated discontents will break out, triggering massive group events. In reality, active participation of migrants can help optimize the migration program. Participation itself is also an interaction and a balancing of benefits between the migration agencies and the migrants. This will greatly increase the migrants' recognition of the project. It will also help mitigate their feeling marginalized and vulnerable, increase their courage and confidence toward a new life and consequently improve their ability to counteract impoverishment risks.

In social actions against poverty, what we need to address are the poor as "actors". After the beginning of the 1990s, in answer to the transition of the international anti-poverty intervention ideology, the introduction of the "participatory development" concept and the discussions on the development subjects and the subjectivity of the poor triggered by this concept can be seen as a step beyond the income-based "abstract poor" concept in anti-poverty intervention practice. People began to pay attention to the understandings, decisions, and actions of the "concrete poor" toward their own social situations.[7] Notably, sociological studies on the "concrete poor" during this period came to the same realization almost at the same time: to emphasize the intrinsic connections between countering poverty and reconstructing communities, and between self-organization of communities and self-development abilities of communities.[8] The subjectivity of the poor does not simply mean the "self" of the subject-object dichotomy in the pure sense of economic rational people, but also includes the insertion of the poor into the communities. As Zhu Xiaoyang has noted, "These poor people are not to be saved through a surgery-like poverty alleviation approach. Their exchange activities or economic activities are inserted into the social relations of their communities. In a way, their responsibility is also based on the social

[6] Wei (2010).

[7] Li (2001a, 2001b).

[8] Shen (1997).

relations and values where they are rooted".[9,10] From the inter-subjectivity point of view, the rationality of the poor can only be construed in a given community and a give context. It can only be integrated and developed in a given community and a given context, too. The process of poverty alleviation should be one that guides the poor communities to participate and construct their communities. The subjectivity of the poor is the key factor for the success of a poverty alleviation policy. Also, the participation of the poor itself should become an institutionalized goal of poverty alleviation.[11]

References

Bao, Z. (2006). The definition, category and some other issues about ecological migration. *Journal of Central University of Nationalities*, no. 1.
Cernea, M. M., & Guggenheim S. E. (Eds.). (1993). Anthropological approaches to resettlement: Policy, practice, and theory. Westview Press.
Li, X. (2001a). Poverty alleviation and governance in China: Analysis of targeting and transfer of counter-poverty resources. *Jilin University Journal*, no. 4 (2015)
Li, X. (2001b). Participatory development. China Agricultural University Press.
Ni, Y. (2007). The theoretical analysis and empirical method to the West China, *Ecological Economy*, no. 2.
Shen, H. (1997). Poverty alleviation transfer and community self-organization. *Sociological Studies*, no. 1.
Wei, S. (2010). On the sustainable development and redevelopment of non-voluntary migration. Wuhan University Press.
Xu, B., & Wang, H. (2000). Development illusions. Central Compilation & Translation Press.
Zhu, X. (2004). Principle of charity and the agreement between tom and jerry. *Open Times*, no. 6.

[9] Zhu (2004).
[10] .
[11] Shen (1997).

Chapter 8
Nonfarm Employment, Labor Dispatch, and Informal Labor

Jieying Zhang

To the ecological migrants, the transformation of life does not simply mean a geographical relocation. Accompanying changed residing spaces are changed livelihoods, too. When they move from the southern mountainous areas up to the north, the farmers have to change their farming habits step by step to adapt to a vastly different climate, landform, and geographical space. More importantly, the supportive or poverty alleviation policies implemented in conjunction with the migration policies are greatly changing the labor models and livelihoods of the migrants. More specifically, on the one hand, stimulated by poverty alleviation through industry, host villages try to introduce a range of industries capable of increasing the farmers' incomes—from scale livestock farms, cash crop farms to solar farms, and even manufacturing plants. Development of these industries helps increase the informal job opportunities for the farmers. On the other hand, the town or village governments are also active in encouraging the farmers to work outside to increase their incomes and reduce poverty. They try finding more opportunities and channels for the employment of the farmers in various direct or indirect forms. Besides, in labor migrant villages, as programmed, these labor migrants will be fully separated from agriculture and make a living by working outside.

Equally important is the fact that the land status in the relocated villages is far different than in the original place. Generally, the land in a migrant village is in one of the following statuses. The first relates to early migrant villages, such as those prior to the 12th FYP period. In these villages, the per capita land area is more abundant and the villagers are in a better position to engage in planting or breeding. Yet they are also faced with difficulties such as poor soil quality, or even year-to-year degradation (such as desertification or salinization), and increasing water unavailability. All these make it increasingly difficult to feed the family by farming alone. The second relates to new migrant villages. These villages are mostly faced with small per capita land

J. Zhang (✉)
Institute of Sociology, Chinese Academy of Social Sciences, Beijing, China
e-mail: zhangjieying@cass.org.cn

X. Wang (ed.), *Ecological Migration and Targeted Poverty Alleviation in Ningxia*, International Research on Poverty Reduction,
https://doi.org/10.1007/978-981-19-7888-3_8

area and extremely centralized land across the village. They would mostly prefer to circulate their land collectively to enterprises. In this situation, the farmers only know how large their land is but have no way to find out where exactly their land is located. While receiving the proceeds from circulation, the farmers have also lost the chance of cultivating the land themselves. This compels the farmers to do nonfarm work to maintain subsistence. The third relates to labor migrant villages. These villages are specifically allocated with families having young labor powers. As programmed, the villagers are directly brought "upstairs". That is, they are directly settled in buildings in a residential neighborhood and are therefore not engaged in farm work anymore. Whichever the case, changed land status poses a propelling force for the migrants to pick up nonfarm jobs.

Not only has the land status changed. After relocation, the geographical spaces hosting the migrants have also changed. Compared with the original place in the southern mountainous areas, the host place where the villages are located usually has convenient transportation and is close to the urban districts. The spatial change also makes it possible for the migrants to find a job in the urban area. Finally, accompanying the changed residing place are changed economic lives of the farmers. In the host place, on the one hand, the above-mentioned changes in land status have made it almost impossible for the farmers to meet the basic needs of life, such as eating and drinking, in a self-sufficient manner. On the other hand, as the supply of consumables is more convenient and abundant, if resettled in buildings, the migrants have to get used to paying for water and electricity supply and for sanitation service. These will all cause them to rely more on purchases to meet their needs. Their demand for cash will greatly increase accordingly. This also poses a driving force for the farmers to work outside and earn instant incomes.

In short, the succession of pushing and pulling forces from policies, land ownership, geological residing space, and own consumption needs compound to make the nonfarm employment of ecological migrants a prominent phenomenon. Increasing shares of nonfarm income in farmers' total income are also a foreseeable trend.

So far, nonfarm employment is already very common. Few families with young labor powers would remain purely self-employed. Incomes from nonfarm employment also constitute the main, if not the primary, source of income for many families. According to statistics, within the villages surveyed, after relocation, over 60% of the families rely on nonfarm employment as the main source of income, over 80% have their members working outside, and around 50% have at least two members working outside. Around a half of the interviewees had the experience of working outside during the previous year. Among these workers, nearly 70% have their income from working outside accounting for more than a half of the total family income. In reality, considering that the time of our door-to-door interview was mostly working hours and some of the outside workers may have been working away from home, the number of families and persons engaging in nonfarm employment in the migrant populations and the proportions of their income are actually both greater than indicated by our door-to-door survey.

Then, what are the characteristics of nonfarm employment of the ecological migrants in Ningxia? What are the opportunities and challenges? Behind the figures, what are the experiences and concerns of the farmers in their nonfarm employment, and what are the difficulties and risks facing them? In this chapter, we will try to answer these questions through qualitative studies to: on the one hand, render a picture of nonfarm employment; on the other hand, try to probe into the characteristics and problems related to nonfarm employment of farmers and further reflect on how the ecological migration cases of Ningxia would help us understand poverty eradication and urbanization. In other words, How does the nonfarm employment of ecological migrants relate to poverty alleviation? What are the similarities and differences between the nonfarm employment of ecological migrants and general farmers working in cities? Does it represent part of local urbanization? Data used in this chapter are quantitative data from an August 2016 survey of seven host villages in Ningxia. Selection of these seven villages covers the north and central parts in geography and involves general ecological migrant villages, labor migrant villages, and old and new migrant villages in category. Data collection was made mainly by means of focus groups and interviews with village officials and farmers, assisted by observation of the villages, farmer families, and industry parks or factories. Besides, the research was also supplemented with some quantitative data based on a large-scale door-to-door questionnaire jointly conducted by the Institute of Sociology of Chinese Academy of Social Sciences and Beifang University of Nationalities.

8.1 Characteristics of Ecological Migration and Nonfarm Employment

8.1.1 Overview of Nonfarm Employment: New Opportunities for Differential Employment and Diversification

The employment of ecological migrants varies grossly by age and gender. For men between 18 and 50, construction is a big area of employment, with available jobs covering almost all tasks available on a construction site: from high-technique electrician to carpenter, bricklayer, plasterer, and bar placer. Of course, doing construction does not occur only after migration. In reality, men had already had the tradition of working experience in the construction field before relocation. Many workers mentioned having more than ten years of working in the construction industry and finding work by virtue of a relationship network built over the years. Nevertheless, relocation has affected men's employment in the construction industry at least in the following two ways. First, relocation has brought more local job opportunities, especially in the host place. Accompanying the relocation of migrants is usually a round of municipal constructions and industry park constructions, including the installation of roads and plantings. These construction projects have offered job opportunities for the local residents. Furthermore, the host place is geographically more advantageous. From the capital city's Yinchuan down to the county seat and towns, the transportation is more convenient. With these conditions, finding a local job will mean being able to live at home. This not only saves the living cost, but also allows the workers to attend their own farm work back at home. An example is Yuanlong Village of Minning Town. According to the village official, most of the local construction workers are working within the town or at least within the autonomous region. As he put it, "Men would earn 120 yuan per day if we chose to work in Yinchuan. That is no better than earning 80 yuan a day just out of door". Some villagers say that they used to work in Yinchuan or Wuzhong before, but after migration, they mainly work in town. They agree that relocation does provide more local job opportunities. However, there also are villagers who do not think that relocation has made much difference to their employment. Villagers thinking this way had mostly worked in other provinces like Shaanxi, Henan, Qinghai, or Inner Mongolia for many years before migration. They usually have worked longer years with rich experience and do skilled jobs. Presumably, relocation of affected populations should not be much a matter for those working outside the autonomous region. But to those working within province, it offers more local job opportunities. This presumption, however, still pends further verification with statistical data.

Unlike men, women are more engaged in more flexible informal work in addition to their own farm work and house chores at home. A typical job is fruit picking. Fruit need a great deal of manual work during the picking season. This offers temporary job opportunities for women. Every year from May to July, labor companies will organize the women to pick wolfberries in the neighboring Qinghai Province, for a period of 40 days to two months. The women are paid according to the weight of the fruit picked. The price is generally 5–6 yuan per kilo. A skillful woman can pick more than 50 kilos a day and can earn more than 10,000 yuan at the end of the time. To the locals, picking is a task of skill rather than strength. Women have clever hands and quick minds. They work faster, earn more and are more suitable. In reality, as this type of work is seasonal, short term and time-flexible, it is more suitable for women who have to attend family work at the same time. In the local area, when it is the picking season, older members and kids will also join in the army of the pickers on a part-time basis to help support the family. According to the villagers, "Even the kids can earn some 20 or 30 yuan a day each". Relocation of affected populations coupled with poverty alleviation through industry initiatives has brought more job opportunities for women. The crop and animal farming industry parks put up near the villages, for instance, are providing more picking jobs. The wolfberry and red raspberry farming base near the villages has become a new destination of local part-time employment for women and older members of the family. Take the red raspberry base of Binhe Jiayuan in Yueyahu District, Yinchuan as an example. There, the village has collectively circulated 800 mu land to a company. The raspberry base can employ over 1000 people in the high season. Fruit picking is paid on a tray basis. According to the company operator, a good picker can earn as much as 100–120 yuan a day. However, as we were told by an old man working in the field, he had been working from early in the morning till sometime after 3 o'clock in the afternoon without a meal. Now he had picked 8 trays of raspberry. And the can pick 17 trays at most. This way, his daily income is less than 60 yuan.

For both men and women, the industries near their villages have brought more job opportunities. Apart from agriculture-rated enterprises, there are more industrial enterprises such as cashmere factories and textile factories. As a village official of Miaomiaohu Village, Pingluo County noted, among the 1400-plus households of the village, there are more than 2200 nonfarm workers. Among them, more than 1800 work constantly and stably. In the peak season, 400–500 people will work in a "factory out of door". Most of the young adults work in the factories. Their number even exceeds that of workers in the construction industry.

Xinmin Community in Xinghai Town of Dawukou Longhu lies adjacent to the Pingluo Industry Park area. There, Hengda Cotton Textile Factory alone offers over 800 job opportunities. Among the current workers there, more than 100 are ecological migrants. The textile factory mostly employs women workers and pays them 2000–3000 yuan per month. For the 7000-plus labor powers available in the village, the opportunities are still limited. Reassuringly, the factory is planning to expand. In addition to increasing the number of jobs, it has also relaxed the employment standards. For instance, the age limit has been broadened from younger than 35 to younger than 45. The health standard has also been broadened so as to provide more job opportunities for the migrants. Similar examples also include Binhu Jiayuan in Yueyahu, which has introduced red raspberry farming, meat rabbit farming, cashmere factories, and clothing factories to resolve poverty alleviation. Among the 802 rural families of this village, there are 378 registered and tagged impoverished families. As planned, a third of these families will have been lifted out of poverty by the end of 2016. So far, over 40 families are working in the red raspberry park and these families can also employ more villagers to join them during the picking season; over 20 households are working in the meat rabbit farming park; over 20 people are working in the cashmere factory; over 30 people have be enrolled into the clothing factory.

Apart from these jobs, the migrants have actually found a range of part-time jobs in the proximity of their new place of residence. These include handling red brick, cooking for the factories, doing services in the business district, cleaning up, embroidery, suturing work in the cashmere factory, pollination or grafting on the farms, as well as planting trees or weeding in planting projects, and even stripping garlic alone. These jobs are mostly short-term, part-time tasks. Yet to the migrants, they are also an important way of earning cash incomes. As a farmer explained, "As long as you are hardworking and find the right job, you can earn 80–90 yuan a day. With that you get a bag of wheat flour".

8.1.2 Characteristics of Nonfarm Employment: Localization, Part-Time Work, and Informal Labor

One of the characteristics of the employment of ecological migrants in Ningxia lies in the division of labor: Men work more stably, while older members and women work flexibly and attend agricultural labor at home at the same time. Besides, whatever the gender and age, they are mostly engaged in seasonal, informal work. Compared with general farmers in China, they are more closely connected to both their former homes and agriculture. Except labor migrants who were not allocated with land from the very beginning, and villagers who have no land to cultivate as their land has been collectively circulated, most of the farmers are trying to deal with both animal and crop farming at home and working outside. Those who work constantly in the construction industry are "migratory birds". They work outside for 6 or 7 months a year and then come back home for the long winter to pass. In reality, the majority of the migrant workers in the construction industry of China are "migratory birds". They work outside in normal times and come back home during the Spring Festival or busy farming season. However, by comparison, the construction workers in Ningxia spend shorter times working outside and longer times staying home in winter.

Working locally provides more chances of attending agriculture at the same time. The size of family farming, including the area of land cultivated, the number of cows or goats raised, and the type of crops grown (this is because different crops require different labor input: For instance, growing corn does not need labor-intensive, continuous attendance. Hence working outside is possible) are all constrained by the available labor powers of the family. Each family will try to settle the division of labor between working outside and farming at home and their proportions according to their own labor availability. Related to part-time work is the means of labor power reproduction. Unlike the "splitting" of working outside—as their incomes are insufficient to pay for the reproduction of their families, the migrant workers have to work in the host area, while labor power reproduction and family parenting is carried out back at home—For locally working migrants living in their own homes, no "splitting would be needed".[1]

The type of work and form of employment of ecological migrants are also unique in themselves. As discussed above, apart from the construction industry and general industries or services, many rural households are actually employed in the agricultural sector. Although they earn wages, as they work in the planting or breeding parks near their villages, they are engaged in agricultural labor, such as seasonal jobs like picking and pollination. This differs significantly from general migrants working in cities. In Yuanlong Village, for instance, 70% of the labor powers are working for agriculture-related enterprises like farms and vineyards near their villages. Besides, as stated above, doing part-time jobs is very common. Neither the locally working men nor the majority of the older members and women are in constant employment. They are in a way casual workers.[2] Firstly, working outside is seasonal, generally lasting from April to October or, at the latest, November of the year. Few would try to find a job in the wintertime. Secondly, there is a great flexibility in the type, content, and time of work. Frequently changing place and content of work is normal for part-time jobs.

[1] Burawoy (1976), Ren and Pan (2006), Shen (2006).

[2] Zhou (2007).

This seasonal, flexible way of employment to which the locals are accustomed has greatly influenced the local government's understanding of stable employment and unemployment. A village official believed: "What is stable employment? To us, being able to find a job for 7 or 8 months a year means stable employment". Being able to work for even 4 or 5 months a year can also be regarded as having work to do. "Being able to find a job for a dozen days a month" is just equal to "having work to do". This understanding of "stable employment" is quite common, whether it be officials or ordinary villagers. This understanding also shows that the villagers are very flexible in their employment expectations and have an extremely high tolerance to unemployment. In recent years, the economy is turning down and the number of job opportunities is decreasing. Quite a few villagers say that before, they could work more than 20 days a month, but now they can only work 10-plus days. However, to them, this is at best an instance of "hard to find work". Closely related to the part-time form of employment is the employment approach of labor dispatch. Regarding this regime, a more substantial analysis will be performed in the following text.

8.1.3 Labor Dispatch System

Migrants are employed mainly in two forms. In the past, farmers mainly relied on their personal connections or employment information communicated through an informal network of relations to find a job. This is particularly evident in the construction industry. Labor dispatch, in which a labor dispatch company organizes the villagers to work in an enterprise, is a new form of employment for casual workers. It is playing an increasingly important role in post-migration nonfarm employment.

In the local area, labor dispatch companies are run either privately or by the village committee. Currently, private operation is the principal form. Generally, private companies charge larger sums of money from the workers, while companies operated by village committees are to some extent non-profiting. They only charge traffic and labor costs. The function of these companies is mainly to build a tie between the needing enterprise and the villagers. Their operation model is: The company negotiates with the enterprise to know about their labor requirements. Then, it recruits the required number of workers from the village and transports them to the enterprises to work there.

Labor dispatch companies need to walk and talk to win job opportunities from enterprises. Competitions exist among different companies. In terms of winning businesses, the private dispatch companies have richer experience and relations. The village collective-run labor dispatch companies have their unique advantages, too. For instance, a village collective is more able to negotiate with the enterprise that uses the village land and win more job opportunities for their villagers. When circulating land or introducing enterprises, they can use providing jobs for their villagers as an additional condition.

Workers are recruited every day. Every morning, villagers desiring to work outside are gathered on the empty ground of the village. The bus of the labor company arrives right on time. After recruiting an adequate number of hands, it transports them to the company to the enterprise to work there. A village official made a vivid description of this: "Every morning, the farmers start off carrying water and steamed bum with them. They arrive at the square merrily, hoping to get some work to do. The big bus of the labor company comes at the appointed time and calls them to get on. Once you get on the bus, you have the day's work opportunity in hand. If you aren't called or you don't get on the bus, all you have to do is go back home down in the dumps". During the first few days of work, wages are paid on a daily basis. The workers use the first few days to familiarize themselves with the work. The enterprise will select more satisfying workers within these days. "If they work honestly, we keep their phone number; if they are lazy, there's no job for them". In a few days, this round of employment relatively stabilizes. The company will give the workers a workpoints card to record their daily workpoints. The wages are paid every 10–20 days.

To the farmers, labor dispatch is something like a "double-edged sword". On the one hand, undoubtedly, the labor companies share off part of the proceeds from the labor of the villagers. On the other hand, they provide information and traffic for the farmers and make their employment channels smoother. Besides, the labor dispatch companies work as a middleman mitigating potential wage default. Enterprises pay remunerations on a monthly basis. They do not pay wages until end of the month. To workers recommended by a labor dispatch company, some enterprises would pay the farmers in advance. In reality, labor dispatch can both jeopardize the interests of the farmers and take a commission on their incomes, and also play a positive part in protecting the farmers and alleviating poverty. The drawbacks and potentials of this regime will be further discussed in the following text.

8.2 Risks and Challenges Associated with Nonfarm Employment of Ecological Migrants

8.2.1 Inadequacy of Job Opportunities

On the one hand, income from nonfarm employment is becoming increasingly more important for the migrant families. Yet it has to be noted that falling labor prices and decreasing job opportunities have also been very common since 2015. According to statistics of Dawuhou Longhu, in 2015–2016, the labor income dropped by an average of over 1000 yuan per person per month from 2013 to 2014, the reduction rate was around 40%. High-wage enterprises, for instance, used to pay more than 4000 yuan per month before. But now they are paying 3000-plus yuan. In the service industry, those who used to earn 3000-plus yuan are earning around 2500 yuan now. In the construction industry, according to data provided by the villagers, in 2013–2014, they used to work an average of 20 or more days per month, but in 2015–2016, they are "out of work half a month". Accompanying the "hard to get a job" phenomenon is the falling daily wage. Before, skilled workers used to earn 280 yuan per day. In 2015, this reduced to 240 yuan; in 2016, it dropped to 220 or even 100-plus yuan. Accordingly, the daily wage of unskilled workers has dropped from 120 to 80 yuan. The same thing has happened to agricultural employment, too: "The wage's down". An example is wolfberry picking. In 2015, the price was 7.5 yuan per kilo; the villagers could earn around 90 yuan per day each. In 2016, the price was 12 yuan per kilo; the daily income dropped to around 70 yuan.

An intuitive perception commonly noted by the villagers is "more people, fewer jobs, less money". Behind this phenomenon, there are more profound reasons. The first is the structural aspect of economy. Economy is turning down. Construction and industry are shrinking. These constitute an essential cause behind decreasing job opportunities. Second, as migrants move in, this round of municipal constructions is drawing to an end. For instance, roads, plantings, as well as plazas and commercial streets are being completed one after another. The job opportunities related to these constructions also decreasing accordingly. On the one hand, job opportunities are decreasing. Decrease of job opportunities, on the one hand, and continuous inflow of migrants, on the other hand, compound to increase the number of surplus labor powers. Competition is posed among the migrants. This causes the labor price to fall further down.

Village poverty alleviation through industry projects has the potential of attracting labor powers. Examples include the light textile plant in Dawukou Longhu, the solar power enterprise, farm and husbandry company and vìneyard in Yuanlon Village of Minning Town, and the red raspberry farming park and meat rabbit farming base in Binhe Jiayuan Third Village of Yueyahu. The projects introduced into the villages have truly created new job opportunities for the migrants. These industries can offer labor prices roughly comparable to the market price, helping resolve the employment of the villagers. Also, to the villagers themselves, "working out of door" is undoubtedly more preferable to working away from home. So far, introducing industries has truly increased job opportunities. Yet as the industry poverty alleviation projects have not run for a long time, and many of them have not even operated for one year, how they will develop and what risks they will encounter still remains to be seen. One of the risks is, although all industries have received a variety of policy support at present, the heavily promoted poverty alleviation measures will not persist. In future, in the absence of vigorous policy support, these industries will be faced with market risks and competitions. Whether they will be able to survive and continue to provide stable job opportunities still remains to be seen.

In reality, parallel to decreasing job opportunities and falling labor prices is the recessive unemployment of the farmers. As discussed above, the understanding of both the local villagers and governments about "employment" and "stable employment" is quite broad and flexible. For instance, working continuously for around a half of the year and being able to work at least a half of the month are both considered a "stable employment". In a way, their tolerance to unemployment is fairly high. Yet, from another angle, this "stable employment" as understood by the locals is essentially a "recessive unemployment". When other incomes of a family, such as agricultural income and policy compensations and subsidies, are relatively adequate and the family simply uses employment income as a supplement, recessive employment will not matter much. But, as relocation draws to an end, the poverty alleviation work comes to a conclusion and the agricultural income continues to decrease (to farmers no longer engaged in agricultural labor, this is almost unescapable), underemployment will prevent labor power reproduction and even give rise to impoverished families, especially at times when families need large amounts of cash, such as when children go to college or old members fall ill.

In short, nonfarm employment has become an increasingly important source of income for the migrants. Accordingly, stable employment is becoming increasingly important, too. Job opportunities are largely determined by the macroeconomic conditions and other structural aspects. Yet poverty alleviation policies are time-dependent. In the long run, how to continuously and stably provide and even increase job opportunities and ensure that the migrants are stably and substantially employed will be a great challenge facing migration projects.

8.2.2 Absence of Labor Rights Protection and Labor Security Under the Labor Dispatch Regime

It may appear that labor dispatch is a bridge between enterprises and workers allowing an effective allocation of labor resources. Yet to the villagers, employment on labor dispatch also poses potential issues. Under labor dispatch, migrant workers cannot receive the same pay as regular workers doing the same work. Nor do they have access to the benefits and guarantees they are entitled to. First, flexible as it looks, labor dispatch is actually paying the least for labor power reproduction. To enterprises, the laborers are simply "at their beck and call". When they need more hands, they add employees. When their sales or orders reduce or when the ongoing process requires fewer hand, they reduce the number of employees. By employing the most suitable workers when needed, they avoid the need of "feeding" a constant number of workers as other enterprises do. This is undoubtedly a means of minimizing cost and maximizing profit. To the workers, labor dispatch is a day-to-day job and they are paid for the days worked. That is, they have no income if they don't work. In labor dispatch, being "left idle" when there is no work makes unemployment recessive.[3] Second, unlike regular employment, the workers are not entitled to paid off days like regular employees, let alone sick leaves or other benefits. Additionally, enterprises are not obligated to pay for the employees' labor insurance and social benefits. This part of the cost is actually passed on to the local government and the laborers themselves.

Compared with their regular counterparts, workers on labor dispatch not only lack necessary welfare, but also lack the associated labor security, especially with regard to employment safety and work-related injuries. Regarding this problem, the labor dispatch companies are short of substantial recognition at current and their counteractions are fairly simple. For instance, a man in charge of a labor company

[3] Biao (2011).

explained to us that they prevent safety hazards in this way: At the time of recruitment, they "make clear to the villagers in the beginning that they will bear all responsibility themselves. Just inform the boss in advance if you are ill. If you are ill today, just don't go there... Tell the boss your problem anytime. Neither of us is liable to the other". In reality, this "make clear in the beginning" and "neither is liable to the other" are not legally binding. Nor can they replace the mandatory obligations of labor security. In the current labor dispatch regime, the employment relationship is unclear. No contract has been signed. The workers do not have employment injury insurance. They are hardly able to protect their rights and claim for compensation against the enterprises, either. Although accident casualty rarely happens, for the migrant workers and their families, any such incidence could be fatal. It would not only pull the family back to poverty, but also pose heavier burden to the local government. In our interview with households receiving precision-targeting poverty alleviation, we discovered that many of the families are impoverished because their primary labor powers are disabled and have lost the ability to work due to work-related injuries or occupational diseases incurred in their previous work history.

The current labor dispatch can yet be regarded all a flexible, effective way of employment. At present, it agrees perfectly with the flexible employment form and part-time working habit of the migrants, allowing the effective allocation between labor resources and opportunities at the highest efficiency. These should never be denied. In reality, if a family relies mainly on agriculture, has other stable sources of income and works part-time only to supplement family expenses, labor dispatch will be very helpful for the flexible employment of the farmers. But if a family relies mainly on nonfarm employment and their employment is mainly dependent on labor dispatch, an employment and social security system should be established and maintained to further resolve the benefits and security of the laborers. Depending on informal employment or working part-time with instances of "being left idle" every now and then as a main means and long-term plan of feeding the family will pose austere challenge to both the farmers and the local social security system. As to how to optimize the labor dispatch regime, a more detailed discussion will be performed in the following text.

8.2.3 Predicaments Facing the Employment of Special Populations

The greatest challenge facing the nonfarm employment of ecological migrants is the employment of laborers over the age of 50. The people between 50 and 65 are not seniors having lost the ability to work. Especially in agriculture, the most familiar area to the migrants, these people can even serve as the main labor power for the family. Even older people wouldn't find it a problem to support themselves by doing some small-sized planting or breeding. Yet labor powers over the age of 50 are not welcomed in the nonfarm market. Almost all villagers and local officials acknowledge that people over 50 can hardly find a job in the job market. Only a few can find work as a cleaner or gatekeeper. For this, they are paid from a few hundred to a thousand yuan per month, which is far lower than regular workers. Even so, these job opportunities are still rare. Some men over 50 have to join women and children in short-term picking tasks. Yet they are uncompetitive even in this type of work. Their income is far lower than the women's for the same duration of time. For a migrant family that has land to plant or space to breed, this should not be a serious problem. For a migrant family whose per capita land area is limited or whose land has already been collectively circulated, the inability of the middle-aged or older members to find job can cause increased family burden and further exacerbate poverty. The appearance of a population able to work but not having a job opportunity can generate even more social problems to the local area. Admittedly, the lack of a job opportunity for the middle-aged or older population is associated with the scarcity of job opportunities, it is also related to the above-mentioned labor dispatch regime. Labor dispatch always selects the strongest and obedient honest laborers. They would not give a chance to those who are not old enough to retire but are slightly less competitive.

Also, suffering are disadvantaged populations like the disabled. If in an area of nonfarm employment, they are more likely to find out ways to feed themselves, at least at a low level of livelihood, by planting or breeding. If in a migration where their land is scarce or has been collectively circulated, they will find it hard to make a living as they can no longer remain self-sufficient by farming or find opportunities in the nonfarm market.

Another group faced with employment obstacles are the ethnic populations. The inability of the Hui people to find a job is partly associated with the stereotyped impressions about this population commonly found in the local area. One is that the Hui people are "lazy"; the other is that they are "hardly manageable, troublemaking". These stereotyped impressions formed at the time when working outside first appeared, the market was extremely short of standardization and the workers had no experience of working outside. As a village official who himself is Hui recalled, "…stayed there for 20 days, nitpicked about the wage and conditions, asked for an unreasonable amount of wage, refused to work and left when rejected, so that (the employer) had to compromise with more compensation. A few people left a bad impression about us Hui people". Yet a closer analysis would find out that, on the one hand, as this village official noted, they are "a few" and cannot represent all Hui workers; on the other hand, there are more essential causes behind the "troublemaking". As another Hui village official explained, if the wages of two months were not paid, the Hui workers would ask to get paid. If their request were not satisfied, they would immediately leave or even "fight physically or orally". Once they did this, they would rally and act in group. So, they could never work long. Through analysis, we will discover that this does not indicate that the Hui workers have in themselves anything intrinsic and unchangeable. In reality, in the construction industry where labor-capital disputes take place most often, wage default is almost a normal phenomenon. In Ningxia, it is customary to pay wages of a few months or even half a year together. Refusal of the Hui workers to tolerate wage default or their distrust to the employers is not a problem of their own. It is only because they are more likely to combine and support each other, and to express their appeals by taking collective actions. In reality, in the construction industry, workers are sometimes forced to protect their entitlements in a "violent" manner. This is not a special case unique to the Hui people of Ningxia, but an outcome of the normalization of multilevel subcontracting, multilevel payments on account, and multilevel wage default.[4] Hence, if a better employment system is established that safeguards the benefits of workers and provides lawful, non-violent collective negotiation channels, the possibility of "troublemaking" will be greatly minimized.

In reality, lack of job opportunities for the disadvantaged populations on the labor market is almost a common problem among all labor markets. However, for ecological migrants, there are some other "disadvantaged" ones: They are disadvantaged not because of objective conditions such as age or ability to work, but because as "newcomers", they do not have the necessary information, experience and relation network. For some villages, if the village itself does not have good conditions with which to negotiate with enterprises, or if the village used to rely on farming before with no history of working outside and without an experienced, socially connected leader and able man, it will be very hard to get job opportunities for the villagers. Some villagers have no working experience themselves, yet their means of existence has changed from farming to working. They have little sources of information or social connections in the host place. Hence, finding a job has become a problem. In a labor migrant village, for instance, the village itself is unable to find employment channels for the villagers. The villagers as new settlers do not have the necessary experience and information, either. Except 40% of the families working constantly in other places or provinces, the remaining 60% are still all highly under-employed. The villagers claim that they are "idle most of the time". As they lack social connections and barging abilities, the average wage of the village is even slightly lower than the other villages. Generally, the wage of working part-time is 100 yuan per day for men and 70 yuan per day for women, which is slightly lower than the average of the other villages. Besides, they have to do even lower-end jobs that other workers would not want to do and their jobs are even less skilled. As they do not have an experienced, socially connected and capable leader, they are more likely to worry about "not getting paid". Obviously, in a context of economic downturn and universal scarcity of job opportunities, "newcomers" are more likely to become disadvantaged in competition and are more prone to unemployment. Additionally, the roles of the village collective and leader are extremely important. An experienced able man can provide social connections and information for the villagers. A capable and resourced village collective can introduce projects, negotiate with enterprises over land circulation and bring more job opportunities for the villagers. If migrants have no collective to support them, no able man to lead them, and little experience and information, they will easily be disadvantaged in the job market.

[4] Ngai and Huilin (2010).

8.2.4 *Myths About the "Quality" and "Wait, Depend, Want" Discourses*

In the previous subsection, we have analyzed the structural and institutional aspects of the nonfarm employment of ecological migrants. In the next subsection, we will try to discuss two discourses frequently heard in both local contexts and government texts, and academic studies—the criticisms of migrants as being "low quality" and "wait, depend, want". These discourses look at personal problems as a major reason behind the poverty of migrants. In discussing how to encourage migrants to overcome poverty and achieve prosperity and how to promote employment, there often exists a very common belief: The migrants' quality is "too low" to meet the requirements of the job market; they are too "lazy" or "afraid to work" to adapt to the intensity and rhythm of working outside; their "wait, dependent, want" conceptions pose a key obstacle for overcoming poverty and achieving prosperity. Admittedly, this observation has some realistic basis. But the problem is: First, it personalizes structural problems like poverty and development. Second, it simplifies the complicated history and reality into a stereotyped impression, a popular discourse and even a "stigma". Continuing to repeat and emphasize these discourses will do no good to poverty reduction in its true sense.

In reality, the availability of job opportunities is determined by the market and the state. In other words, this depends on the general economical conditions, as well as the concrete policy options and regulatory measures. If job opportunities are under-provided relative to labor availability, there are always some labor powers unemployed or recessively unemployed. However, laborious or hardworking they are, the overall availability of job opportunities will not increase. Conversely, if these people get themselves employed, another more disadvantaged group of laborers will be elbowed out of the labor market. Regarding farmers in the southern mountainous areas, a popular claim is that because they have the history of "living by what they can find", being "lazy" and "undisciplined" has already become a deep-rooted habit. A migrant who himself is a village official argued that this is totally a "misinformation". He responded this way: "Mountain people are more tolerant to hardship. Why? Because conditions in the mountains are poor and they invest more hard work in the year". In another village, the village official also had the habit of using being "lazy" to explain poverty. In reality, after further inquiry, we discovered that the so-called lazy is usually underlain by a more profound background and complicated reasons. Here are some examples. Some village official explained that the Hui men are "lazy" because they "are more fond of animal farming than working outside". This indicates that this phenomenon is more related to their different working habit and method. It is also associated with the current scarcity of job opportunities. Another village official explained the "laziness" of the villagers this way: They "refuse to borrow loans and reject project funds; they rely on two members working outside and refuse to raise cattle". In this example, "two members working outside and don't want to raise cattle" becomes an evidence of being "lazy" again. In actual fact, families may have their own considerations when they refuse to borrow loans or do not want to participate in poverty alleviation programs. Another village official noted that some villagers are "lazy": "They have no older members to support. Their children can go to school. Their house has been put up. They can also work outside. They have no worry about their life over the year. That's enough. There's no need to take action". In this claim, the villagers accused of being "lazy" are actually working outside. They are "lazy" because they are satisfied with their economic situation and do not want to seek further development. From these examples, each accused instance of being "lazy" has their own reasons behind them.

Similar to being "lazy" is the claim that the migrants are "backward-minded". For instance, an official noted that some young women in the village "(insist) looking after their children at home. They could otherwise be really good labor powers. These young, good labor powers are exactly what is wanted (by the enterprises)... Yet they insist coming back home;" however "it takes time and patient guide to change idea... If they are working couple, they earn 5000 to 6000 yuan a month. That's enough for a living". In this example, the women are "backward-minded" not because of the subjective problem of the "mind", but the objective conditions. Women shoulder the family responsibility of parenting and looking after children. They can't afford to work out and do house chores at the same time. If young women are to work outside, that would be no more than passing on the responsibility of parenting children to the older members and creating new "left-behind children". Some impoverished families have provided such realistic examples. This family has three children. The youngest is three. Last year, in order to supplement the family income, the wife went to pick wolfberries in Qinghai for two months. The children were left with the husband's father during this time. This year, as they can no longer count on the old man for help, the wife has to give up working in Qinghai. In another family, the husband works as a technician at an electricity plant in Qinghai. The wife works in the cashmere factory and looks after two children at the same time. She gets up at 5 o'clock every morning and gets everything ready for the children for the day before she can leave for work. The older girl of the two children is responsible for cooking lunch and looking after her brother. This arrangement is all right solely because of their extra working hours and the older daughter who shouldered part of the parenting responsibility. Nevertheless, different families have different conditions. Not all women can use the same arrangement for themselves.

As to the claims that the migrants are hard to manage, don't work long, or even "easily offended", there are actually complicated reasons. As they have just transitioned from agriculture to a nonfarm employment area, it is not surprising that the migrants are not adapted to the rhythm and labor regime of working outside yet. Besides, the migrants are not very comfortable with the enterprise' default of their wages. They want to get paid as soon as possible, do not trust the employer, yet have little necessary experience, information or negotiation channels. What they can do is but "create a trouble" or choose to leave before the end of the work period. As a village official remarked, "It's not that we don't work. We'll get used over time". This village official also explained that the chemical enterprise nearby would default their wages for 3 or 4 months in normal times, and sometimes hold the wages unpaid for 5 or months, but "when they are used to doing that, give in to it, do not want to leave, they will be willing to keep working there".

Similar to "low quality" are the "wait, depend, want" discourses that are quite popular among local and official texts and academic discussions. By "wait, depend, want", it criticizes that farmers would not try to overcome poverty and achieve prosperity on their own initiative, but would wait for, rely on or even ask for the welfare, assistance, and compensation made available by various government policies. In these discourses, farmers are described as people who seek development passively without depending on their own abilities. But if farmers are not simply regarded as an object of criticism but a rational actor engaged in agricultural production, what they can do is but take actions in the visible opportunity space to win available resources and maximize their benefits. In other words, "wait, dependent, want" is precisely the choice for the villagers in a visible, achievable opportunity space and institutional environment to maximize their family income. In a way, they are actually not passive. Instead, trying to benefit from policies is the outcome of rationally balancing and planning livelihood in an active manner. The actions of farmers are actually closely related to policy environment. As such, to resolve the villagers' policy dependence, it is better to optimize the policy design and ensure an equal interest allocation mechanism to better utilize the subjective initiative and better motivate the enthusiasm of the villagers, rather than simply accusing them of "wait, depend, want".

The above-mentioned "quality", and "wait, depend, want" discourses are extremely common. Of course, we also admit that the migrants were not borne skilled, eligible, compliant workers conforming to the enterprises' requirements and accustomed to the labor regime. In history, any transition from agricultural populations to industrial labor powers has taken a long transitioning and even making process.[5] In reality, it has not been easy in any industrialization or urbanization process to turn self-employed small farmers into employed workers. In other words, the "quality of an eligible worker is not innate, but takes time and even generations to complete. The exception is that "quality", as a discourse unique to China,[6] personalizes the under-supply of job opportunities in the process of macroscopic socioeconomic transformations or the structural mismatch between labor powers and opportunities. These criticisms on individuals are not beneficial to promoting employment, increasing income, or reducing poverty. Nor are they good for the decision makers to make appropriate preparations and policy arrangements. Additionally, the seemingly being "lazy" and "passive" on the surface usually has more complicated historical and social sources. Only by abandoning criticisms simply objectifying the migrants and closely distinguishing these reasons is it possible to truly effectively motivate the migrants to overcome poverty and achieve prosperity more independently and more actively.

---

[5] Thompson (1963).

[6] Yan (2008).

8.3 Conclusions and Policy Responses: Ecological Migrants Among Migrant Workers

8.3.1 Conclusions

So far, very few migrant families are not engaged in some form of nonfarm employment. Nonfarm employment has become an important source of inform for the vast majority of the ecological migrant families. As long as there are labor powers, there will be long- or short-term formal or informal employment in the migrant families.

The nonfarm employment of the ecological migrants in Ningxia is found to have the following distinct characteristics. First, there is a conspicuous work division between genders. Men are more engaged in long-term work while women mainly work part-time. Men are more engaged in the construction industry while women are engaged in a variety of jobs. Among them, seasonal picking is a typical job for women. These jobs are mostly available in the agriculture-related enterprises on the periphery of the villages, like crop and animal farming parks. Second, jobs are seasonal, flexible, and "casual". The working time of even the men believed to be in "long, stable employment" is no more than seven months a year. The employment form and timeline for women and older members are more flexible. This way of taking part-time jobs enable the farmers to both work outside and farm at home. Unlike the "split" form of labor power reproduction for typical farmers, this form of employment gives a substantial flexibility to the understanding of "stable employment" and a relatively high tolerance to unemployment among the locals. Third, accompanying part-time and casual employment is the labor dispatch form of employment. Labor dispatch companies privately run or run by village collectives send the casual workers to the needing factories every day, building a bridge between enterprises and farmers. Fourth, this round of poverty alleviation policies, especially the industry-driven poverty alleviation projects, has brought more local job opportunities for the migrants. Among these enterprises, the proportion of agriculture-related projects is extremely high. Although they work on employment, the migrants can still engage in agriculture there. They are responsible for part of the work in the planting or breeding process either as contractor or as worker.

Nonfarm employment of ecological migrants, as an increasingly noticeable reality, both carries the hopes and opportunities of the villagers and farmers for overcoming poverty and achieving prosperity, and is faced with appreciable predicaments and challenges. First, to the local governments, the greatest challenge is undoubtedly how to create job opportunities so that most of the interested and able migrants can work stably and substantially. According to information provided by the villagers and officials, the employment situation over the recent couple of years has not been optimistic, with trends of decreased opportunities and falling income. Recently, in the policy background of precision-targeting poverty alleviation, some villages have introduced some projects in an attempt to drive development through industry. These projects have indeed provided opportunities for the employment of the villagers. Yet, when this round of poverty alleviation projects comes to a conclusion, in the absence of policy support, whether these industries can still provide stable job opportunities for the villagers still remains to be seen. Second, the seemingly flexible, efficient labor dispatch regime also entails related risks. The most essential problem is the failure to fully account for the rights and benefits of the workers and the inability to ensure that the labor powers are substantially protected. In this situation, for a family relying mainly on income from nonfarm employment, once unemployment or a work-related injury occurs, the possibility of being impoverished again is very high. Third, one predicament in promoting nonfarm employment of migrants is the difficulty for the less informed, less experienced and less socially connected "newcomer" populations to obtain a job opportunity. This is particularly true for the middle-aged and older populations between 50 and 60. Although they have the ability and willingness to work, they are not welcomed in the labor market. Labor idleness of these populations both poses a major obstacle for families to get rid of poverty and increase income and is also negative to the harmony and stability of the local society.

Finally, criticisms of the villagers as being "low quality", and "wait, dependent, want" exist in both local contexts and academic studies, regarding these problems as the source of poverty of the migrants. In this chapter, we have analyzed some of the discourses, and pointed out that only by overstepping these problem-personalizing, farmer-objectifying criticisms can we identify the sources of poverty and obstacles for poverty alleviation according to facts, and adopt policy designs that give better play to the subjective initiative of the migrants on this basis.

Except part of the labor migrants—in the policy design, this part of the labor migrants are simultaneously relocated and urbanized and changed into wage laborers, most of the ecological migrants still remain land-connected. Yet, what they face are not merely changes in the natural environment, but also changes in geographical location, land ownership and use, and a new policy environment. When they relocate, urbanization more or less takes place, too. All these will compel the migrants to rearrange the input and work division of family labor power between agricultural and nonfarm labor to maintain their subsistence. They are no longer pure agricultural laborers and are different from typical migrant workers, too. They are between agriculture and industry, farmer and worker, and even farmer and migrant worker.

8.3.2 *Policy Responses*

8.3.2.1 Precisely Analyze Unemployment Reasons, Provide Targeted Employment Services

This chapter holds that similar to poverty alleviation projects, service and support actions for promoting the employment of migrants should also be precise. First, targeted promotion of employment depends on a more accurate or subtle mastery of the reasons of under-employment on the macroscopic dimensions. For instance, rapid response and readjustment should be made to structural unemployment, namely, the mismatch between labor supply and industry demand. Second, an important thing is to precisely identify the reasons that cause the unemployment of the migrant individuals and families, and provide differentiated job service schemes pertinent to different reasons and different populations. For the disadvantaged populations in the job market, local governments should negotiate with the introduced enterprises to create more guaranteed job opportunities for them. For instance, regarding the employment of disabled people, Binhe Jiayuan has managed to convince the meat rabbit farm to mainly employ the disabled people of the village by negotiating with the introduced enterprise. This is a good example to follow. In addition to guaranteeing job opportunities for the migrants, the local government should also analyze the characteristics of different populations and guide them into industries where they are more advantaged. For instance, labor powers above 50 may not be suitable for new industrial labor, but they have richer experience in agricultural labor, and can be more included into agriculture-related enterprises.

Finally, being precision also means the need to overstep the "quality", "wait, spend, want" discourses, and stop taking the migrants as lazy and backward objects or recipients of criticisms or moral education. Under this concept, first, it is extremely crucial to provide the migrants with more practically useful skills in line with the reality of the job market. In our survey, 21.6% of the migrants have participated in government-organized job training. Yet only 40.5% of these migrants think that these training are helpful or highly helpful. Obviously, in terms of providing more effective, non-formalized job training, there is still large room for further improvement. Second, it is also important to precisely analyze the reasons behind being "lazy", "backward-minded" and "hard to manage" and solve these difficulties. For instance, providing substantial child-rearing services will allow the young women to become more worry free when working outside. Providing a good, friendly working environment for the stigmatized populations, especially a smooth collective negotiation approach, will free them from fears, barriers, and worries. The violent form of the locals in protecting their rights has left an impression of being hard to manage. For this reason, some building projects would rather employ migrant workers from other places and even workers from the south. Regarding this situation, a more formal collective labor-capital negotiation mechanism should be established to both protect the rights of the workers and eliminate the distrust of the laborers to their employers, as well as the misunderstanding of the employers to the local people.

8.3.2.2 Develop Non-Profiting Labor Dispatch Services, Expedite the Establishment of a Labor Security System

As discussed above, regarding the current labor dispatch regime: On the one hand, it has indeed interlinked enterprises and workers more effectively; on the other hand, there is a problem of under-provision of labor rights and guarantees. To solve this problem, one effective solution is: The township and village collectives play a more active part in the labor dispatch regime. In the case of a government labor dispatch organization, this organization is more able to negotiate when facing enterprises and more able to win trust when facing villagers. According to our survey, more than a half (54.9%) of the villagers would like to obtain jobs through government introduction rather than through labor dispatch companies (4%).

The above-mentioned labor dispatch company established by the village collective of Yuanlong Village, and the village collective labor dispatch company of Xinmin Community of Xinghai Town, Dawukou Longhu, both have the potentials of playing a greater role in protecting the villagers and negotiating collectively. We were told that the company of Yuanlong Village only charges the traffic and organization costs. They try to minimize the commission on the villagers' income, and would pay the villagers wages on behalf of the enterprises when they cannot make the payment on time. In reality, a labor dispatch organization led by the township or village collective can go even further in non-profiting services. First, by defining a labor dispatch agency as a "social enterprise" or a "non-profiting organization", it ensures that this company is different from other private enterprises and does not share profits with the villagers. Besides, this organization can also shoulder more collective negotiation functions and help the villagers protect their own entitlements. They can also negotiate on behalf of the villagers in the event of an accident or dispute. A labor dispatch company of the village collective is not advantaged for remote job opportunities. But when nearby enterprises are concerned, the village collective does have the resources and ability to negotiate. Why trying to win job opportunities for the villagers, it can also protect their entitlements. For example, an idea of Dawukou Longhu company is: They will oversee the contract signing between enterprises and the villagers and oversee the performance of the contracts. In the event of wage default, the company will help the villagers ask for payment from the enterprise. In the event of work-related injuries or other labor disputes, it will participate in the negotiation or provide legal assistance for the villagers. In the event of other problems, they will also be willing to help them solve these problems. These ideas are all worth promoting. If a non-profiting labor dispatch company can really perform these functions, its most evident effect will be keeping the villagers undamaged and protecting them from being impoverished again as a result of injuries. In the long run, this is also good for the normalized and positive development of the nonfarm employment of the migrants.

Besides, nonfarm employment of ecological migrants has become their main source of income. Whether or not anticipated, the nonfarm trend of migrants is already a reality. As discussed above, the migrants between farmer and worker have changed or are changing into employed laborers. Parallel to this reality are the different degrees of social security burden to the host place. Predictably, as migrants continue to move in and age, this burden will continue to increase, too. In this situation, it becomes extremely important to include labor security into consideration and build a preliminary labor dispatch system. On the one hand, having substantial labor security means providing a security network for the migrants, greatly reducing the risks of impoverishment caused by diseases, injuries, or unemployment. On the other hand, this also greatly relieves the local social security burden and contributes to a more normalized, positive social security system.

To summarize, migrate relocation projects will generally consider the environmental carrying capacities of the host place, including the population carrying capacities of land and water resources. Our survey has discovered that there are instances of water scarcity and land under-irrigation in the host places of ecological migration. In reality, this is true with labor employment, too. Undoubtedly, nonfarm employment has already become an important source of income for the migrants. Then, for migrant resettlement, ensuring job opportunities is equally important as ensuring environmental resources. Regarding the nonfarm employment of ecological migrants, a dynamic, long-term, structured insight is particularly needed. While relocating, the migrants are also involved in the wave of agricultural modernization, industrialization, and urbanization in one way or another. Hence, a more forward-looking program is needed. In face of increasing labor supply in the host place, we cannot simply rely on transferring these labor powers to the east and south. Instead, we should know that parallel to disappearing demographic dividends and the labor shortage in the east and south are new situations like economic new normal, industry restructuring and upgrading, machine-for-man, and even machine-for-man in agriculture, as well as the inhumanity facet of taking "split" employment as an expedient (causing families to separate, resulting in numerous left-behind old people and children). Hence, we should, in light of the local advantages and characteristics, further explore into the forms of local employment and farmer–worker mixture in the local area, develop distinctive agricultural areas and absorb more labor powers. Of course, although informal labor between farming and working outside is conducive to the effective allocation of labor powers and the maximization of family income, and also agrees with the current labor and living habits of the local people, when exploring this form of employment, labor rights and benefit protection cannot be absent. Instead, we should not leave "informal labor" in an ungoverned area. We should plan in a coordinated and forward-looking manner and intensify the construction of formal labor security. In short, resettlement is not the end of a migration project but a new start. We should allow the migrants to have a chance to maintain subsistence, overcome poverty and achieve prosperity through their own labor. We should also allow them to work with dignity and decent, protect them against unemployment risks and prevent the formation of new poor populations.

References

Biao, X. (2011) Global "body shopping": An Indian labor system in the information technology industry. Princeton University Press.

Burawoy, M. (1976). The functions and reproduction of migrant labor: Comparative material from Southern Africa and the United States. *American Journal of Sociology, 81*(10), 1050–1087.

Ngai, P., & Huilin, L. (2010). A culture of violence: The labor subcontracting system and collective actions by construction workers in post-socialist China. *The China Journal, 64*, 143–158.

Ren, Y., & Pan, Y. (2006). The spatial politics of transnational labor process: Dormitory labor regime in the age of globalization. *Sociological Studies, 4*, 21–33.

Shen, Y. (2006). The social transformation and reformation of Chinese working class. *Sociological Studies, 21*(2), 13–36.

Thompson, E. P. (1963). *The making of the English working class*. Pantheon Books.

Yan, H. (2008). New masters, new servants: Migration, development, and women workers in China. Duke University Press.

Zhou, D. (2007). The "free city edge people": A study on the casual workers in the southeast coast area of China, Sun Yat-sen University Press.

Chapter 9
Ecological Migration and Social Governance

Xia Li

Implementing ecological migration is a major strategic step of the party committee and government of Ningxia Hui Autonomous Region toward coordinating urban–rural development and building a harmonious socialist society. It is an important practice of deepening the West Development Initiative and an important approach for benefiting special disadvantaged populations and helping them get rid of poverty and become better off. In essence, the ecological migration project of Ningxia has lived up to the basic spirit of the five-in-one construction objective (i.e., economy, politics, society, culture, and ecological civilization) in the Report of the 18th National Congress of the Communist Party of China. That is, to relocate populations in areas unsuitable for human production and living through the construction of political and social management institutional mechanisms aimed at economic-ecological civilization and cultural progress, promote the transfer of impoverished populations from the southern mountainous areas of Ningxia to the Yellow-River Basin economic area, raise the production and living levels of the migrants, radically resolve the poverty problems in the southern mountainous areas, protect the national ecological security, and realize sustainable development. Ecological migration, as a way, a means, or an approach of social governance, is an important content of reflecting ecological civilization and beautiful China initiative and innovating the social governance system and mechanism in the new era. During the implementation of ecological migration, the party committee and government of Ningxia Hui Autonomous Region have actively explored the road of social governance and greatly promoted the rapid socioeconomic development of the autonomous region.

X. Li (✉)
Ningxia Academy of Social Sciences, Yinchuan, China
e-mail: nxskylx@163.com

X. Wang (ed.), *Ecological Migration and Targeted Poverty Alleviation in Ningxia*, International Research on Poverty Reduction,
https://doi.org/10.1007/978-981-19-7888-3_9

9.1 Problems and Challenges Facing Ecological Migration in A New Economic Normal

As China enters into an economic new normal of "a slowdown in economic growth, making difficult structural adjustments, and absorbing the effects of previous economic stimulus policies," the existing social structure has been broken. The interest pattern has become more diverse and complicated. A series of contradictions and conflicts have arisen among different interest groups when fighting for and protecting their own interests, exposing ecological migration to problems and challenges.

9.1.1 Noticeable Problems with Increasing Migrants' Income

After settling in the migrant villages, compared with their original place, the migrants are enjoying much better housing, electricity supply and travel convenience. Yet the mere farmland allocated at two mu per person and irrigated on water supply rationing is far less than enough for becoming well-off. The available path for migrants to increase their income is quite narrow yet.

(1) Prominent supply–demand imbalance of water resources in the new migrant resettlement destinations

The ecological migrant resettlement destinations are mostly located in new reclamation areas. There, the soil is infertile with poor land establishing conditions and is costly and time-consuming to develop or modify. The migrant settlements are far from irrigation canals and have no water use quota. The supply–demand imbalance of water resources is very prominent. Monopolizing and stealing water have occurred very frequently. Due to irrigation problems, there have been a steady succession of water-related disputes. Competition for water resources is already threatening the stability of the migrant society.

(2) Limited self-development abilities of the migrants

Having lived deep in the ill-informed mountains, the migrants are mostly inferior in both the degree of education and labor skills. Even fewer are proficient in a particular line. This, coupled with the lack of related effective training, makes labor output and transfer a tough task. It is highly difficult to rely on planting or breeding to become better off. The self-development abilities of the migrants are quite weak. "Moving out" is not a problem, but how to "hold up" and "become better off" is a challenge.

9.1.2 Urgent Problems with Governance Service for Migrants

In Ningxia, migration exists in multiple forms and no uniform migration management system is in place. This poses great problems to social governance in the relocation destinations.

(1) Labor migration problems require urgent solution

According to our survey, the majority of the migrants would prefer resettlement with land and are not very interested in labor migration. The reason is that the ecological migration policy is more favorable than the labor migration policy. Under the ecological migration policy, they are provided with one mu of land per person and a 54-m^2 dwelling per household. For the house, the government provides a subsidy of 25,000 yuan per household and the ownership rests with farmers. Under the labor migration policy, they only have a 40-m^2 turnover dwelling without ownership and can only live there on rent. Comparison of realistic benefits has caused the vast majority of the migrants to prefer ecological migration. As we were told by the poverty alleviation office of Yuanzhou District, Guyuan City, less than 15% of the people in Yuanzhou District would prefer labor migration. In Longde County, among the 10,002 people relocated in 2011, 8100 people or 81% would prefer ecological migration. Those preferring labor migration made up only 19%. It was hard to fulfill the labor migration task assigned by the autonomous region.

(2) Social governance problems of ecological migration are particularly prominent

First, management of new ecological migrant villages.

During the 12th FYP period, among the 34 built-up new ecological migrant villages, except 12 that are villages of the original inhabitants in their entirety, the remaining 22 new ecological migrant villages are mostly 3000 or 4000 people or even 5000 people in size. During the relocation, a lot of manpower, material, money, and energy were invested by both the government and the individuals, especially the migrants, who relocated with their whole family with hopes and expectations for future. After relocation, as the villages are as large as thousands of people and the migrants come from different villages and even different townships, it takes time for them to integrate into one another. Some of the migrant villages are already becoming unstable. Some villages have even had violations of law and disciplines. Some migrants have chosen to appeal to the higher authorities for help.

Second, lag in land reallocation.

As far as we have learned from the survey, among the built-up migrant villages, less than a half of the land has been truly allocated to the villagers in some counties. In some counties, the migrant villages have been occupied more than a year, but the land has not been realloted to the villagers yet. With no land to cultivate, the migrants have responded strongly.

9.1.3 Industrial Development Problems

As the relocated migrants are impoverished ones from the mountains, affected by economic hardship and traditional concepts, they commonly lack fund and skills. Nor do they have the initiative to develop industries. Follow-up industry development is an important support for the migrants to become better off. It is an important guarantee for the long-term development and income increase of the migrants. Nevertheless, as far as we have learned from the survey, crop farming by the migrants in some counties has not broken away from the old stereotypes of cultivation. Water-saving agriculture or facility agriculture is just a sporadic "potted landscape" but not a mass of flowers. Even those small arch sheds assigned among the migrant families are not cultivated or operated at a high level. The performance is less than good. Animal farming is performed in small quantities and poor qualities. Large as a migrant village is with over a thousand households and thousands of inhabitants, there are few households raising cattle. As to labor output, more are engaged in physical, low-income labor. Fewer are in skilled, high-income occupations. To date, the dominant industry for the migrants is still agriculture. A large part of the population is deposited inside agriculture. This not only results in water resource shortage, but also limits the potential of the migrants to further increase their income.

9.1.4 Worsening Soil Salinization

Hongsipu migrant area is the largest ecological migration destination in China. Since salinization began in some sections in 1999, salinization in the irrigation district has already covered 580.5 ha today, making up 25.5% of the farmland of Hongsipu irrigation district. The consequent abandoned land area is 184.53 ha, accounting for 31.8% of the salinized land area. The salinized land is principally located in three controlled irrigation districts, covering Dahe Township, Shaquan Township, and Baosibao Town. Intensifying soil salinization has hampered the sustainable development of Hongsipu.

9.2 Basic Framework and Practice of Social Governance of Ecological Migration

9.2.1 Basic Guideline for Social Governance of Ecological Migration

Social governance is a human-oriented governance. Based on multi-subject cooperation and subjective participation and guided by scientific, standardized rules and systems, it aims to better cope with social problems, enhance the rational allocation of social resources and meet the reasonable requirements of the people. The social governance of ecological migration is an important part of modernizing the national governance system and governance capacity, and an important guarantee for maintaining peace and content for the people, stability and good order for the society, and lasting security for the country. The fifth plenary session of the 18th CPC central committee has emphasized that "We need to build a safe China, improve the social governance system led by the party committee, dominated by the government, collaborated by the society, participated by the public and guaranteed by the rule of law, make social governance more precise, and build a social governance pattern by the people and for the people." The "Recommendations of the CPC Central Committee for the 13th Five-Year Plan for Economic and Social Development" has made all action plans for strengthening and innovating social governance pertinent to the prominent problems in the area of social governance. It is of great significance to advancing the modernization of the national governance system and governance capacity. Hence, in innovating social governance, we should emphatically grasp and deal with several relationships between moral reconstruction and the improvement and perfecting of the social infrastructural institution, between democratic construction and legal construction, between transparency and democratic election, and between the decentralization of administration and the stimulation of social vitality, and advance the modernization of the national governance system and governance capacity. Specifically, the social governance of ecological migration in Ningxia should follow the basic guidelines below.

(1) Enhance service

We should highlight the service concept, focus on the improvement of people's lives, raise the management level in service, achieve the organic unity between social governance and public service, vigorously resolve the most-concerned, most direct, most realistic interest problems of the migrants, and allow them to share the outcome of reform and development.

(2) Seize the core, enhance the construction of rural grassroots party organizations

We should practically seize the core of building rural grassroots party organizations, uphold the principle of focusing on the central task and serving the overall situation, take active actions and resolve practical problems.

(3) Highlight legal thinking and methods of rule of law

We should strengthen comprehensive rectification at source and application of policy, address both the symptoms and root causes, highlight legal thinking and methods of rule of law, push the social governance of ecological migration toward the front end and precision, raise the social governance level of ecological migration in all aspects, and provide strong guarantee for the stability of reform and development across the autonomous region.

(4) Establish an effective inter-linkage system

Governments at all levels should encourage communities, social organizations, and social workers to participate in the ecological migration governance and service, build an inter-linkage between the government and communities, social organizations and social workers, enhance mutual trust, bear own responsibilities, achieve positive interaction between government governance and social regulation and resident autonomy, and effectively resolve various problems.

(5) Make great efforts in building rural spiritual civilization

We should practically do well in building rural spiritual civilization as a vehicle, actively cultivate the practice of the core socialist values, vigorously optimize the human settlement, perfect cultural service, endeavor to raise the civilization level in the rural society, build a healthy pattern of jointly advancing rural grassroots governance, and provide strong, effective guarantee for building a moderately prosperous society in all respects and winning the battle against poverty in pace with other parts of the region and the country.

9.2.2 Exploration and Practice of Social Governance of Ecological Migration in Ningxia

Strengthening and innovating social governance represents an inevitable choice for China to cope with social problems and realize healthy and stable socioeconomic development. It is an important duty assigned by the party and the people to party committees and governments at all levels under the new historical conditions. During the ecological migration process, party committees and governments of Ningxia at all levels have, centering on strengthening the construction of rural communities, cultivating and developing social organizations and comprehensive governance of social security, actively explored new approaches for the social governance of ecological migration, gradually built a new institutional mechanism of grassroots social governance, safeguarded the overall situation of the society, and effectively promoted the rapid socioeconomic development in the autonomous region.

9.2.2.1 Perfect Public Service By Highlighting the people's Livelihood

(1) Strengthen public through comprehensive rectification of rural environment

Since the implementation of comprehensive rural environment rectification in 2008, the autonomous region has spent an aggregate of 2189.5 million yuan (including 600 million yuan of central government fund) on the comprehensive environmental rectification of all administrative villages in the region. They have built 128 central domestic pollutant treatment facilities, 1450 km of sewage pipes, 285 rubbish transfer stations (points) and landfills and over 17,000 trash bins; distributed

7845 garbage collecting and transportation trucks; protected 406 rural central water-head sites; and created 195 national and autonomous region-level ecological townships/towns and ecological villages. The rural environmental outlook has greatly improved, benefiting all the rural populations in the region.

(2) Improve the lives and capabilities of the disabled

Since 2012, the autonomous region has continuously implemented home accessibility projects for impoverished disabled people. This well-being program involves installing removable kitchen tops, voice-activated rice cookers, toilet handles, and voice intercom doorbells as needed by the seeing, hearing and physically impaired. According to statistics of the Ningxia Disabled Person's Federation, in 2016 alone, they installed accessibility facilities for 400 disabled persons for free, helping improve the lives and capability of the disabled people.

(3) Helping female ecological migrants start their own business

To improve the status of rural women, the first thing to do is to improve the economic status of women. The autonomous region government encourages the women migrants to start up their own business and provides small amount guaranteed financial discount loans to women migrants who start up their own business in animal farming, agritainment, agricultural and sideline foods, or cultural products. From January 2012 to December 2016, the autonomous region government granted 242 million yuan of small amount loans and supported 4416 rural women to start their own business.

(4) Attach great importance to the aging problems of ecological migrants

With the circulation of rural land, young people have been leaving to seek subsistence in cities. The number of elderly who live by themselves has been rising year after year. The autonomous region government has continuously explored new approaches for rural elderly care service and encouraged private enterprises to energetically develop rural elderly care services. In 2014, Tongli Village of Litong Town, Helan County set up the first village elderly care service center. Its 94 one-story rooms and 106 beds have effectively resolved the elderly care in and near the village. Elderly who live by themselves can enjoy life care, cultural entertainment, and health care just out of door.

9.2.2.2 Continuously Innovate the Model of Ecological Migration Governance

Recently, Ningxia has continuously innovated the model of ecological migration governance, having conducted full-scale standardization of township (subdistrict) livelihood service centers and comprehensive governance service centers. Shizhuishan Citizen Livelihood Service Center has set up 55 public service items under six categories. In 2015, the center handled affairs for more than 500,000 people, and served another 40,000 people by attending their first petition letters or visits, providing rights protection service and offering legal education. It has become an "all outpatient clinic" safeguarding the harmony and stability in the grassroots. In 2015, thirty-nine volunteer service lawyers across the city participated in the mediation of petition matters. They handled 1219 legal assistance cases for free for people in difficulties and recovered economic loss of 40.57 million yuan. Shizhuishan City was the recipient of the 2015 "Best Case of National Social Governance Innovation" award jointly sponsored by People's Daily, people.com.cn, and National School of Administration.

9.2.2.3 Push the Ethnic and Religious Affairs onto the Track of Rule of Law

Ningxia has a Hui population of more than 2.3 million, which is more than a third of the total population. Hence, management of ethnic affairs holds a decisive significance across the region. In order to push the ethnic and religious affairs of the autonomous region onto the track of rule of law, the region has promulgated 162 local laws and regulations around the "Law on Regional Ethnic Autonomy," having safeguarded the regional ethnic autonomy and the entitlements of the ethnic populations and realized the harmony and stability of Ningxia.

9.2.2.4 Practically Strengthen the Construction of Public Security and Prevention Networks in Ecological Migration Destinations

In the ecological migration destinations, the autonomous region consistently carries out volunteer patrol and defense activities like "one head per five (ten) households," inter-household joined defense, and mosque joined defense. It has established mass self-defense, autonomy organizations like "rural security associations," and vigorously promoted harmonious alarm bell, infrared alarm apparatus, and anti-theft bolt for vegetable greenhouses, creating a new landscape of mutual-benefit, multi-win rural modern security control work. As of November 11, 2016, the number of filed eight categories of criminal cases, including self-initiated manslaughter and intentional injury, had dropped by 32% from the same period of the previous year.

9.2.2.5 Vigorously Push Forward the "Investigate and Relieve" Work for Contradictions and Disputes in Ecological Migration Destinations

First, they have explored and perfected the "multipartite mediation" work pattern. In the migrant villages, a special people mediation organization has been established. A "multipartite mediation system" widely participated by all walks of life and involving the inter-linkage among people mediation, administrative mediation, and judicial mediation has been structured.

Second, they have perfected and bring into effect an "investigate and relieve" work system. In ecological migrant villages, a "regular investigation, regular reporting, regular circulation, coordination conference, tag and urge to handle" work system for disputes has been established. In order to relieve multipartite contradictions and disputes, Qingtongxia City has enriched and perfected town and village (community) mediation organizations by grasping the source, symptoms, foundations, and grassroots. Taking the chance of town and village re-election, the number of mediators at various levels has been increased from 638 to 743. The "four-tiers" mediation network of town mediation board, village mediation board, mediation panel, and mediation informant has been seamlessly interconnected.

Third, they have established a religious celebrity-participated mediation work mechanism. In minority-inhabited migration destinations, religious celebrities like prestigious imams who understand the policy or temple management committee directors are engaged to mediate civil disputes of different causes among the religious people as mediator. In Baofeng Town of Pingluo County, the first Muslim "mediation room" in the region has been set up. Young and middle-aged imams with good legal accomplishment and policy levels or temple management committee directors are engaged as mediation volunteers. The people's court of Haiyuan County has formulated the "Implementation Method of Invited Mediators" and invited 11 famous religious celebrities and imams as "invited mediators of the people's court." The successful rate of imam-participated mediation cases is 90% or higher. In Chuangye Community of Hongsipu District, overall the past 10 years, the imam KFH has co-mediated more than 20 ethnic or religious disputes according to law, mediated more than 3000 major difficult disputes and recovered economic loss of more than 30 million yuan. The traffic police group of Tongxin County has engaged 36 imams or temple management committee directors as people's mediators. In their mediation work, the imams and temple management committee directors actively communicate the party's ethnic policy, national law, and religious doctrines. They have effectively relieved large numbers of contradictions and disputes, enhanced ethnic consolidation, promoted family harmony, and played a positive role in advancing the social harmony and stability of Ningxia.

9.2.2.6 Practically Strengthen the Fundamental Construction in Grassroots, Consolidate the Foundation of Social Governance

First, grassroots comprehensive governance organizations have been further improved. As of the end of 2016, all the 238 townships (towns/subdistricts) had had a comprehensive governance committee (office) in place, with 391 full-time comprehensive governance officials. 2731 villages (settlements) had set up a comprehensive governance room integrating the five functions of comprehensive governance, police affairs, security and safety, mediation, and outsiders service and management. The overall level of grassroots comprehensive governance work has been raised. Second, the building of grassroots comprehensive governance teams has been strengthened. Across the region, 123 rural police rooms have been set up, with 131 full-time resident policemen. Full-time joint security defense teams of no less than 10 have been set up in 80% of the townships/towns in river areas and key market towns in mountainous rivers.

9.2.2.7 Lay Special Focus on the Help and Education Management of Special Populations

Thorough search registration and archive handover of inmates are well enforced. Control measures are put into action. The interconnection in help and education management among family, unit, and community is perfected. The actual difficulties of released prisoners in employment, life, and family are actively resolved. Yinchuan City has raised funds and established an entrepreneurship and employment fund for released prisoners and persons under community correction, and identified 10 enterprises and industrial parks as "Yinchuan City entrepreneurship and employment base for persons under community correction or help and education." So far, 71 judicial administrative stations in 4 cities and 10 counties (cities/districts) are doing community correction work. A total of 955 community inmates have been received and no single person has committed a new crime.

9.3 Predicaments and Path Selection for the Social Governance of Ecological Migration

As the ecological migration in Ningxia speeds up, some deep-seated contradictions and problems have become more prominent, greatly affecting the social harmony and stability. Hence, now and for sometime to come, how to continuously innovate the social governance system, strengthen the social governance awareness, raise the social governance level and provide high-quality, efficient social governance service will become an important and urgent realistic topic facing the party committees and governments of Ningxia at all levels.

9.3.1 Predicaments Facing the Social Governance of Ecological Migration

As the economy enters into a new normal, Ningxia is now in an important stage of strategic opportunities, as well as a period of increasing social contradictions. The social governance of ecological migration in Ningxia is showing new situations of diversifying interest groups and intensifying conflict of interest. The previous government-led monotonous "management mindset" is no longer able to resolve the current intricate social governance problems very well. Increased difficulty in the social governance of ecological migration is mainly manifested in the shackle from traditional government management mindset, monotonous governance means, and low-efficiency interest expression.

9.3.1.1 Shackles from Traditional Government Management Mindset

Relevant governments of the autonomous region have not yet to realize how important the social governance of ecological migration is to sustainable socioeconomic development. The mindset of paying too much more attention to economic construction and economic growth than social governance and public service is still very pronounced. More notably, some leading cadres are still only interested in economic development, taking investment promotion and GPD growth as their main work objective and task. Their lack the necessary degree of recognition, research, input, and active promotion of social governance. This is bound to impair the construction of new ecological migrant villages and harmonious communities and the building of harmonious ethnic relations. It will also affect the poverty alleviation work and the rapid socioeconomic development.

9.3.1.2 Monotonous Means of Social Governance, Limited Number and Size of Social Organizations

The social governance of ecological migration is a systematic task that requires the joint participation of governments, social organizations, market, and individuals. Nevertheless, the subject of social governance at present is still the government. As the government has limited capacity and energy, it cannot deal with all social affairs by itself. Hence, this requires social organizations to take an active part and play their function of autonomy and role as a social relief valve. As the number and size of social organizations participating in the social governance of ecological migration, especially those undertaking social governance and public service, are limited, they are far from meeting the actual demand. The monotony of the social governance means will undoubtedly pose many negative impacts on innovating the social governance of ecological migration.

9.3.1.3 Low-Efficiency Interest Expression for Ecological Migrants

First, the farmers' interest expression system has failed to undertake its function of expressing farmers' interest very well. Within the Chinese institution, farmers' interests are expressed through the participation of NPC deputies and CPPCC members in the deliberation and administration of state affairs, the villager autonomy system, and the petition system. Yet these systems have not undertaken their function of

Table 9.1 Comments of ecological migrants in Ningxia about the result of appealing to individuals or organizations for different reasons, percent

Type of dispute	Useful			Useless		Total
	Very	Some	Don't know	Little	None	
Living problems	12.3	30.2	5.6	26.8	25.1	100
Daily routines	13.7	35.3	11.8	21.6	17.6	100
Rights violations	2.8	25.0	2.8	30.6	38.9	100
Administrative executions	5.3	26.3	5.3	31.6	31.6	100

Source The project team based on responses to questionnaire

expressing farmers' interests very well. Second, villager autonomy has not become an effective channel for expressing farmers' interests. As the relationship between village committee and village party branch is not clearly defined, village committees are more functioning as an agency of the township/town people's government instead of an organization truly representing the farmers' interests or reflecting the farmers' interest demands. Hence, the effect of interest expression represents an important factor affecting the interest expression behavior of ecological migrants. Good expression will play a demonstration role that helps improve the initiative of expression and increase the trust in interest expression; low-efficiency or inefficient expression will badly frustrate the interest expression initiative of ecological migrants and lead to their distrust in interest expression. A vicious cycle will form between distrust and interest expression behavior as a consequence. So far, the interest expression of ecological migrants in Ningxia is less than ideal. After expressing their interests without effect for many times, some ecological migrants would choose to go to extremes. This will increase the distrust of the ecological migrants in interest expression and even give rise to the notion "expression is useless." In response to our question "Have you ever appealed to any of the following persons or agencies for life-related problems over the past five years," 65.6% of the respondents chose "No"; 23.3% chose "Village committee"; those ticking "Village party organization," "Township government," "NPC deputy" and "Women's federation" accounted for 1.4%, 1.4%, 0.8%, and 0.2%, respectively. Regarding the expression effect, most of the respondents desiring interest expression believed that the expression was "Useless." More details are presented in Table 9.1.

9.3.1.4 Low Sense of Autonomy of Ecological Migrants About Participating in Social Governance

Farmers are the subject of value for rural social governance. The attitude of farmers is the core criterion for the success of rural social governance. The essential of achieving farmers' recognition is to give play to their autonomy and directly involve farmers into each part of rural governance. The reality of Ningxia is that local governments take care of everything, whether in terms of industry positioning, determination of development plans, or implementation of concrete schemes. Farmers are just lookers on. For this reason, most of the farmers assume that rural governance is a "government business" by nature and has nothing to do with themselves. Some grassroots governments have already "stolen the show" and dissimilated "farmer decides" into "government decides." With the massive transfer of the population into towns and nonfarm industries and the outflows of rural young adults, some of the ecological migrant villages in Ningxia are already hollowed or even empty. The rural social governance subjects are in an embarrassment of "none to elect" or "hard to select." Ecological farmers are absent in rural governance. This is the greatest challenge facing the rural grassroots social governance in the autonomous region.

9.3.2 Path Selection for Social Governance of Ecological Migration

The third plenary session of the 18th Central Committee of the Communist Party of China proposed that the overall goal of deepening reform in all aspects should be to "improve and develop the socialist system with Chinese characteristics, and promote the modernization of the national governance system and capacity." This is a major theoretical innovation of China's socialist construction. Social governance innovation is an important topic for the modernization of the country's governance capacity. On the one hand, social governance innovation requires the government to gradually change their functions, from "versatile" to "limited." On the other hand, governance subjects compatible with the limited government should be cultivated and activated to realize the transfer of government functions. As the role of social governance in improving social management and enhancing grassroots democracy becomes more prominent, giving play to the collaboration and mutual support between government and social organizations in social governance has become the acting point for promoting the social governance of ecological migration at this stage.

9.3.2.1 Raise the Subject Consciousness of Ecological Migrants

First, we should guide the migrants to recognize their status as subject. We should utilize various media to communicate to the migrants on their obligations and responsibilities in rural governance, thereby inspiring their sense of ownership. Second, we should enhance the subject consciousness of the migrants. In concrete rural construction practice, whether in terms of village rectification, industry positioning, or development planning, we should seriously ask for and listen to the opinions of the migrants and continuously enhance their sense of responsibility and mission. Third, we should further improve the villager autonomy system, increase the statutory autonomous performance of village committees, continuously perfect the Organizational Law of Villager Committee, enhance the ability of village committees as the "head" of villagers, play down the role of government as an "agent," and preclude government intervention with the self-governance of villagers. We should practically standardize the villager democratic election, democratic deliberation and village affairs disclosure systems, and ensure that the villagers will exercise their rights of election, democratic decision-making, and supervision properly.

9.3.2.2 Strengthen the Construction of Grassroots Party Organizations, Promote Social Governance Innovations

Grassroots party organizations, as the "bellwether" of rural governance, should work closely around the new requirements of social governance for grassroots party construction in the economic new normal, use strengthening grassroots party construction as a red thread running through social governance, and promote social governance innovation by building grassroots party organizations. First, we should continuously strengthen the construction of village-level organizations, actively cultivate rural leaders for becoming well-off, and provide solid guarantee for ecological migration development. Second, we should improve the village official structure, follow the democracy, openness, equity, best competence principle, inject those well-behaved, educated, capable, responsible, welcomed young people into the village official positions, and give play to the role of grassroots party organizations as the battle base and the role of party members as vanguards and models. Third, we should strengthen training on village officials, raise their qualities by "going out, bringing in" to broaden their horizon, and enhance the ability of village officials to lead farmers toward prosperity. Fourth, we should strengthen the construction of rural spiritual civilization to make grassroots party organizations more vigorous and promote the all-round rural socioeconomic development.

9.3.2.3 Cultivate and Develop Social Organizations, Establish A Multicenter Governance Structure

The process of strengthening and improving rural social governance is also a process of government-society interaction. Hence, we should vigorously cultivate social organizations, transfer government functions in an orderly manner, share part of the social governance functions and social responsibilities with social organizations, and realize "joint governance" between managers and society. First, we should strengthen the internal management of social organizations, effectively raise the internal management level by establishing rules and systems, truly straighten out the social organization management system, and achieve self-management, self-discipline, and self-development. Second, we should intensify supervision and management of social organizations. The government should standardize the rates of charge of social organizations and disclose them to the public at scheduled or unscheduled intervals. It should also advance the construction of a social organization credit system, establish a third-party evaluation system with government guidance, departmental collaboration, and social participation, and make full, objective, impartial evaluation of the integrity of social organizations. Third, we should actively explore and establish a government purchased volunteer service system, set special volunteer service funds, and establish and maintain a volunteer service guarantee and incentive mechanism.

9.3.2.4 Highlight the Roles of Religious Celebrities, Minimize Intensification of Contradictions and Disputes

Considering that Hui ecological migrants make up more than 60% of the total population of the region, we should give full play to the important roles of mosques, imams, and other folk resources among Hui communities. For some of the family disputes, religious disputes, interest disputes among Hui ecological migrants and between Hui and Han migrants, to the extent allowed by law, we can, through imam mediation or by communicating some legal provisions when lecturing on "al-waz," encourage the migrants to try to improve their living environment through hard work to get rich. We should vigorously advocate the good atmosphere of cooperation, friendship, and mutual assistance and promote the building of a harmonious society and relieve social contradictions.

9.3.2.5 Perfect the Interest Expression Mechanism

Perfecting the interest expression system of migrants is an important means of promoting social harmony. It is also an important step toward protecting the entitlements of migrants. To perfect the interest expression system of ecological migrants, we need to properly resolve the interest demands of ecological migrants. First, we should try to effectively link people, administrative and judicial coordination methods together in administrative, policy and legal forms and through guidance, coordination and education means, and resolve different problems in light of the contradictions of different periods. Second, we should implement the "leaders receiving the mass" system, and perfect the responsibility system of petition workers.

9.3.2.6 Innovate Policies, Expedite Legislative Process

The several major social policy innovations in the USA have all tried to resolve social problems through extensive and systematic legislation. For instance, when the 1929 economic crisis engulfed the USA, Roosevelt launched a New Deal and Hundred Days program. In 1933, the USA passed the "Emergency Banking Act," "Federal Emergency Relief Law," "Agricultural Adjustment Law," "National Industrial Recovery Act," and Tennessee Valley Authority Act" at an alarming speed to maintain bank credit, devalue the dollar, stimulate foreign trade, restrict agricultural production to maintain farm product prices, and prescribe association prices to limit competition among businesses. In 1964, Johnson put forward a "Great Society" program to plan, build and beautify cities and resolve pollution and population growth; fight poverty and increase job opportunities; provide social insurance and relief to help low-income families unable to meet their basic needs; support education and open up roads toward a "great society"; build and beautify rural area, exploit depressed areas and built high-speed freeways. He also caused the passage of the elderly medical care system, medical aid system, the Civil Rights Act and the Voting Rights Act. After he was re-elected, he increased the funds for the "Great Society" program, with funds for health, education, and development of depressed areas increasing from 8.1 billion dollars in 1965 to 11.4 billion dollars in 1966. Many of the poor people received social relief. Currently, our urgent task is to strengthen top design, centralize the best expert teams and frontline managers in China to expedite the legislation on civil rights, livelihood and social organization management, establish necessary professional courts and trial procedures, solicit opinions from all sides, and rapidly implement and put in force the related laws and regulations.

Chapter 10
Social Governance of Self-initiated Migration

Yongfang Yang

President Xi Jinping has pointed out, "Eliminating poverty, improving people's livelihood and achieving common prosperity are the essential requirements of socialism", and "The valuable and important points in advancing poverty alleviation development to what we see today are both precision; precision is the key to success". President Xi's speech has expounded the great significance of advancing the tough battle against poverty in China at present. Ningxia, as an underdeveloped region in West China, is among the country's main battlefield against poverty. According to the development realities and the current conditions of the impoverished populations, the autonomous region party committee and government have clearly identified the objectives and roadmap for the region's own tough battle against poverty: The region will strive to lift 58,130,000 impoverished rural residents out of poverty at the current standard, write off all 800 impoverished villages, and uncap all nine impoverished counties and districts by 2018. Within the nine impoverished counties and districts, the per capita disposable income increase rate will be higher than the average of the region and exceed10,000 yuan by 2020; the basic public service will arrive at a level of or close to the national average. For Ningxia to win this arduous battle against poverty, in addition to placing high emphasis on the precisely identified 580,000 impoverished rural residents, attention should also be paid to the part of the autonomously migrated impoverished rural residents, namely the self-initiated migrant population.

In the west regions, self-initiated migrants are usually the minority impoverished populations. Due to religious beliefs, the self-initiated migration problems in these regions are more particular. Petitions and mass events have taken place every now and then, posing great risks to the social stability of the minority-inhabited regions.

Local governments, having too much doubts and worries due to interest game, are far too slow to launch effective policy measures regarding self-initiated migration problems, leaving these programs a hard nut for the social governance at the

Y. Yang (✉)
Ningxia Academy of Social Sciences, Yinchuan, China
e-mail: nxyyf69@163.com

X. Wang (ed.), *Ecological Migration and Targeted Poverty Alleviation in Ningxia*, International Research on Poverty Reduction,
https://doi.org/10.1007/978-981-19-7888-3_10

grassroots government levels. The self-initiated migrant populations in Ningxia share the same characteristics as those in the other parts of the western regions. Recently, the party committee and government of Ningxia Hui Autonomous Region have placed high emphasis on self-initiated migration problems, having launched some measures regarding self-initiated migration problems. Unfortunately, these measures have not been fully brought into effect, leading to increasing numbers of relocated self-initiated migrants and worsening problems. At present, how to view self-initiated migration problems correctly, guarantee the entitlements available for the self-initiated migrants, resolve the sustainable development and outlet problems of the self-initiated migrants so as to keep the self-initiated migrant-inhabited areas under constant security and regulation, and allow the self-initiated migrants to enjoy the same farmer-benefiting policies and equal public service and public products as other enrollees in government-sponsored migration programs is of great realistic importance to ensuring the fulfillment of targeted poverty eradication and the realization of a moderately prosperous society in all respects in Ningxia in pace with the other parts of the country.

10.1 Basic Conditions and Rights Protection of Self-Initiated Migrants in Ningxia

10.1.1 Emergence and Development of Self-initiated Migration in Ningxia

Historically, the western minority regions have long been a more frequently migrated part of China since the ancient times due to the adverse environment and economic poverty there. In order to change the poverty outlook of these regions, since the 1980s, China has stepped up efforts in the poverty governance of the western minority regions. Through poverty alleviation through relocation, whole-village advancement, contiguous development, poverty alleviation through science and technology, poverty alleviation through industry, and poverty alleviation through finance, the poverty status of the western regions has been greatly improved. Ningxia is an underdeveloped western region. Its Xihaigu, known as the "most arduous area under heaven", is one of the 14 closely contiguous destitute areas in China. For many years, the successive party committees and governments of Ningxia Hui Autonomous Region have all placed high emphasis on poverty alleviation development. After a four-stage poverty alleviation campaign—the "Three Xi", "Double Hundred", "Thousand Village", and "Million Population", an aggregate of around 2.9 million impoverished residents has been lifted out of poverty. More notably, during the 12th FYP period, in order to protect the ecological environment and help the impoverished rural populations to get rid of poverty, the government of Ningxia started off another large-scale population relocation in the form of ecological migration, having completed the relocation of 350,000 ecological migrants from the central south. Through ecological migration projects, the impoverished rural population has been greatly reduced. Three hundred key impoverished villages have been wholly written off from the poverty register. The filed and tagged impoverished population dropped from 10,150,000 to 580,000; the poverty incidence dropped from 26.6 to 14.5%; the per capita disposable income increase rate of the farmers in the central south has stayed higher than the region's average for many years. The self-development ability of the impoverished areas has been gradually enhanced. Their production and living conditions have been greatly improved. Their poverty alleviation development work mechanism has been continuously innovated. They have gradually paved out a new road to getting rid of poverty and achieving prosperity for minority regions. In a way, the 12th FYP period marked the five years when the autonomous region has employed the greatest effort in poverty alleviation development, attained the fastest development for the impoverished areas, and earned the greatest benefits for the impoverished populations. Its poverty alleviation development has undergone a historical transition from single-point poverty alleviation to holistic advancement, from broad irrigation to precision drip irrigation, and from fighting alone to collaborated combat. This has laid a solid foundation for the autonomous region to achieve the "Four Ningxia" goal (open Ningxia, rich Ningxia, harmonious Ningxia, beautiful Ningxia) and step into a moderately prosperous society in all respects in pace with the other parts of the country.

Both the "migration through relocation" implemented earlier for poverty alleviation and the "ecological migration" carried out recently for improving the ecological environment are organized and scheduled migrant resettlement forms implemented by the government. We call these migrated people "enrollees in government-sponsored migration programs". For those enrollees in government-sponsored migration programs, as their housing, land, residence registration transfer, and industry development related to whole-family relocation are solely planned by the government, with associated supportive policies as guarantee, they have less pressure and fear about relocation. As policy relocation only covers part of the impoverished population, for the massive impoverished population in the southern mountainous areas of Ningxia, some of them, inspired by the enrollees in government-sponsored migration programs as a model, have no other choice but relocate in endless flows from the southern mountainous areas and central arid land into the northern river area of the autonomous region with their own efforts. These are called "self-initiated migrants". Self-initiated migrants are defined as impoverished rural populations that leave the place of their residence registration through various channels and relocate into an ecologically more favorable area at their own decision and with their own efforts, without relying on the government's organization or enjoying any subsidy from the government, in order to resolve their economic poverty and subsistence problems. Self-initiated migrants in the western regions are characterized by large population, large-scale, and wide distribution. It is safe to say that wherever there are enrollees in government-sponsored migration programs, there are self-initiated migrants. Admittedly, of course, self-initiated migrants are also an inevitable outcome of population mobility. Self-initiated migration helps relieve the ecological environmental stress in the original place and resolve the migrants' own poverty problems on the one hand. On the other hand, due to residence registration problems, the many predicaments facing the self-development of the self-initiated migrants have also brought a lot of social problems to the host place, as well as great pressure to the social governance at the grassroots government levels. The development outlets of the self-initiated migrant population are extremely prominent not only in Ningxia, and these migrants have also become a non-negligible disadvantaged population in the western minority regions.

The emergence of self-initiated migration is in a way the result of the combined action of adverse natural environmental factors and missing management factors. According to R. Herberle's push–pull theory, the widespread self-initiated migration is the outcome of the combined push and pull forces arising out of the tremendous economic, social, and ecological divergences between the original and host places. Self-initiated migrants are usually impoverished populations. Their original places are generally locations with adverse natural environment, insufficient water resources, frequent natural disasters, weak agricultural infrastructure, underdeveloped economy, inconvenient transport, and acute shortage of education and medical resources. As to the host places, in terms of natural conditions, these areas usually possess abundant land and water resources and flat land favorable for agricultural production. Within Ningxia, the host places for self-initiated migration are mainly in the northern river areas, namely the Yellow River-irrigated area. Compared with the southern mountainous areas of the autonomous region, the northern river areas are endowed with abundant land resources, sufficient water resources, plus vast areas of reclaimable, cultivatable deserted, or uninhabited areas. Furthermore, as the terrain is flat, the quality of the reclaimed land is much better than the highland in the southern mountainous areas. The land output effect is quite high. Good natural environment and rich land resources produce a huge "siphon" for self-initiated migration, enticing the self-initiated migrants to come and settle down in the northern river areas with the help of their relatives and friends.

Moreover, the intensifying self-initiated migration problems are also attributable to the lack of government management.

First is the lack of top design and coordinated planning at the autonomous region level. For many years, no consensus has been reached as to whether the autonomous relocation of self-initiated migrants should be encouraged or restricted; no policy planning has been provided for standardization and guidance; and no top design and coordinated arrangement guiding the rational mobility of impoverished rural populations have been produced in light of the novel urbanization progress of Ningxia, the strategic space planning layout of the autonomous region, and the million population poverty relief program. This has left the impoverished rural populations migrating at will and settling down in disorder for many years. Second is the lack of communication and the mutual prevarication between the original and host places, leading to severe absence of necessary government duties and severe absence of service management toward self-initiated migration. The governments of the original places let-go of, and even implicitly consent to or encourage, self-initiated migration. They deem that the outflow of self-initiated migrants into other places will help improve the ecological environment, relieve the supply–demand contradiction between people and land, and also lighten the capital and management pressure of the government. The governments of the host places do not like, do not accept but cannot stop, and cannot do anything about self-initiated migration into their jurisdictions. They are largely afraid that the inflow of large self-initiated migrant populations will affect the local economic development, impede the local urbanization progress and performance indicators like family planning, increase the financial expenditure and fund input related to education, medical care, social security and infrastructure, and add to the management difficulty and work pressure related to public service, social security, population management, and ecological environment. Hence, they have been holding a let-go attitude to the self-initiated migrants buying land or homestead in the host places. Administrative services are absent, and self-initiated migration is constantly kept outside government supervision, giving rise to unorganized, unserviced, and unmanaged self-initiated migrant-inhabited areas such as the "Ximayin Migrant Development Zone" in Xixia District of Yinchuan and the "Xiyanghong Village" in Minning Town of Yongning County. In the absence of government management, inspired by the large-scale government-sponsored migration programs as a model, over the past three decades, self-initiated migrants have snowballed into the northern river areas, producing a colossal army of self-initiated migrants.

Self-initiated migrants are a group of people associated with enrollees in government-sponsored migration programs. Hence, the 30-year government-sponsored migration programs history of Ningxia is also a period when self-initiated migration has taken form. Self-initiated migrant relocation first began in the mid-1980s. It was most concentrated over the 10 years from the 1990s to the beginning of the twenty-first century. More than 90% of the self-initiated migrants come from the impoverished southern mountainous areas of the autonomous region. Recently, with the saturation of the land carrying capacity and the strengthening of grass-roots social governance in the host places, self-initiated migrant relocation has been effectively contained. The number of relocated migrants has been on the fall. According to a special survey made by the Public Security Department of Ningxia Hui Autonomous Region on the self-initiated migrants region-wide, as of February 2015, the autonomous region had 64,384 self-initiated migrant households totaling 231,591 self-initiated migrants. This included 54,817 households, 192,336 residents with intra-regional hukou, accounting for 83.0% of the total migrant population, and 9567 households, 39,255 residents with extra-regional hukou, accounting for 17.0% of the total migrant population. In terms of location, except Xiji and Jingyuan Counties of Guyuan City and Haiyuan County of Zhongwei City, all other counties (cities, districts) have received the relocation of self-initiated migrants. Within the region, more than 90% of the self-initiated migrants come from the mountains in the south or central part of Ningxia; locations outside the region mainly include Gansu, Shaanxi, Inner Mongolia, Xinjiang, Henan, Hebei, and Jiangsu. In terms of age composition, 16,442 of the migrants are between 0 and 6, accounting for 7.1% of the total migrant population; 50,255 are between 6 and 18, accounting for 21.7%; 146,599 are between 18 and 60, accounting for 63.3%; 15,748 are between 60 and 79, accounting for 6.8%; 2547 are over 80, accounting for 1.1%. In terms of residing length, 3705 of the migrants have resided in the host place for less than 1 year, accounting for 1.6% of the total migrant population; 15,285 have resided in the host place for 1–3 years, accounting for 6.6%; 212,601 have resided in the host place for more than 3 years, accounting for 91.8%. In terms of residing status, 28,234 households and 113,942 migrants have their own house and land, accounting for 49.2% of the total migrant population; 28,590 households and 97,963 migrants have their own house but no land, accounting for 42.3%; 3727 households and 8106 migrants have their own land but no house, accounting for 3.5%; 3833 households and 11,580 migrants are residing in a rented house, accounting for 5%. In terms of source of income, the migrants are mainly engaged in agriculture and working outside. 148,681 migrants are engaged in agriculture, accounting for 64.2% of the total migrant population; 49,329 are working outside, accounting for 21.3%; 4631 are private business owners, accounting for 2%; 28,950 rely on other sources of income, accounting for 12.5%.[1] In terms of host place, they are mostly concentrated in Yinchuan, Shizuishan, Wuzhong, and Zhongwei Cities. In Yinchuan, the self-initiated migrants are typically found in the "Ximayin Migrant Development Zone" (it has been taken over by the Xixia District government), Xingjing Town and Zhenbeibao Town of Xixia District, Liangtian Town of Jinfeng District, and Minning Town of Yongning County. In Shizuishan City, the self-initiated migrants mainly reside in Xinghai Town of Dawukou District, Hongyazi Township of Pingluo County, and Hongguozi Town of Huinong District. In Wuzhong City, they are mostly found in Hongsipu District and Sunjiatan of Qingtongxia Town. In Zhongwei City, they mainly reside in Dazhanchang Township, Qukou Farm, Changshantou Farm, and Enhe Town of Zhongning County, and Xuanhe Town of Shapotou District.

Due to the negligent management of the grassroots government in the host place of self-initiated migration, the self-initiated migrants have developed several different residing patterns over the years of migration. Self-initiated migrants residing with enrollees in government-sponsored migration programs in a fill-in form have mainly purchased land and houses from enrollees in government-sponsored migration programs. They mainly engage in planting or breeding at home and work outside in the slack season. In terms of public infrastructure, children's education, and medical care, they can enjoy basically equal treatment as the local villagers. Affected by their residence registration, their endowments, basic living allowances, and medical insurances can only be handled back in their native place. Typical examples include Hongsipu District of Wuzhong City, Minning Town of Yongning County in Yinchuan City, and Dazhanchang Township of Zhongwei City. Relatively independent self-initiated migrant-inhabited areas are typically inhabited by self-initiated migrants engaged in agricultural production on farms subordinate to an agricultural reclamation group or on reclaimed state-owned uncultivated land. They rent or purchase small pieces of barren contracted land from large farm development contractors or farm employees in the reclamation area and mainly rely on crop

[1] Data based on a survey conducted by the Public Security Department of Ningxia Hui Autonomous Region in February 2015, as cited from the "Survey on the Self-initiated Migrants in the Autonomous Region" (internal information).

farming, working outside, and subsidies for returning farmland to forests. Affected by residence registration, they have great difficulty with their children's education. Their endowments, basic living allowances, and medical insurance can only be handled back in their native place. These migrant settlements have long been left ungoverned: The hamlets lack overall planning, the infrastructure construction lags behind, and the farmland ditches, canals, and roads are unmatched. Typical examples include Helanshan Ranch Ximayin Migrant Development Zone and Henan New Village Migrant Settlement in Xixia District, Xiyanghong Village, Shanyuan Company, Lianhu Ninth Brigade Migrant Settlement in Minning Town, and the farm migrant settlement in the Yinxi Ecological Protection Forest project, in Yinchuan City. Overall, the self-initiated migrants vary moderately from one another in terms of economic income, living level, and access to basic public service depending on their host areas, the sector in which they work and the production skills they possess. But the vast majority of them say that they are very contented with their migration and very confident in the future development of themselves and their families.

10.1.2 Rights Protection for Self-initiated Migrants in Ningxia

The western minority regions are areas having the highest incidence of poverty in China. Over the three decades of reform and opening up, the Chinese government has invested heavily in resolving the poverty problems in the western regions. Nevertheless, as these regions are deeply impoverished, quite a large part of the impoverished rural population cannot be covered by government-sponsored migration programs. This part of the impoverished rural population has still managed to change their own poverty through self-reliance, hard work, and independent efforts. Yet, the values and roles of self-initiated migrants have not been objectively realized, acknowledged, or assessed by governments at all levels and the whole society. Their own entitlements cannot be guaranteed. This has not only restricted the sustainable development of the self-initiated migrants, but also greatly affected the management of the government in the host places. The lack of entitlements for the self-initiated migrants is typically reflected in the following three aspects.

First, their economic entitlements are impaired. In order to get settled in the host place, the self-initiated migrants have to purchase or rent land and houses in the host place. For these, each family needs to pay relocation and settlement fee of 100,000–200,000 yuan, which is basically all the money they have. Moreover, they also need to spend money on land modification after moving in. At the beginning of relocation, most of the families would find themselves on a tight budget. Some families have incurred new debts. Development funds are badly inadequate. As they cannot have their residence normally registered in the host place, their land ownership is not acknowledged by the government. They cannot enjoy the various farmer-benefiting policies and government subsidies in the host place. Nor do they have access to microcredit support from financial institutions. This has caused many inconveniences to their lives and production. Some areas even require the self-initiated migrants to pay management fees. Farmers from Yonghe Village, Xuanhe Town, Shapotou District of Zhongwei City told us that they were once charged farming replacement fee of 30 yuan per mu of land by the local government. As this was strongly opposed by the self-initiated migrants, triggering a mass petition, the government had to cancel this charge. As their residence registration is not in the host place, it is impossible for the local government to identify the impoverished households from the self-initiated migrants according to the standard for the filed and tagged households. This has prevented them from enjoying the various poverty alleviation policies, leaving a "dead angle" for targeted poverty relief and overall poverty eradication.

Second, their political status is marginalized. Due to residence registration problems, the self-initiated migrants do not have legal identity in the host place. The local government refuses to acknowledge or accept them, whereas the government in the native place holds a let-go or put-off attitude. The self-initiated migrants have remained unacknowledged, disrespected "unregistered residents" all the time, living in the crevice between the original place and the host place. As we have discovered in our survey of Ningxia, some self-initiated migrants have already lived in the host place for more than 30 years. The first generation of the self-initiated migrants are still somewhat attached to their native land, but the second generation of the self-initiated migrants have totally regarded themselves as native. Their lifestyle, opinions, and interpersonal relationships have already been fully blended into the host place. In the host place, the self-initiated migrants have no right to vote or be voted, no access to the rights available to a regular villager, and no say over issues relating to their own vital interests. They obviously lack an institutionalized channel for interest expression, their wishes are not respected, and their rightful entitlements are not guaranteed. They have fallen into a political situation of collective aphasia.

Third, they lack public service. Affected by residence registration, the self-initiated migrants cannot enjoy equal treatments and guarantees as the host residents. Payment and reimbursement of endowments and medical insurances and residence registration of newborn babies have to be handled back in the native place. After their children have started school, they have to take the high school and college entrance examinations back in the native place. As we were told by the self-initiated migrants from Yonghe Village, Xuanhe Town of Zhongwei City, their children have to attend school more than 5 km away from home every day, escorted by their parents however bad the weather is. They also have to pay 80–1350 yuan as temporary schooling fee or school selecting fee per school term. Nor can they apply for social aids such as basic medical insurance compensation. The self-initiated migrants are virtually isolated outside the current public service system, increasing the living costs for the migrants. Additionally, the self-initiated migrant hamlets lack overall planning, the infrastructure construction lags behind, and the farmland ditches, canals, and roads are unmatched. "Marginalization" of the self-initiated migrants has led to a quite common anxiety among this group of people. Some of the migrants have come up with the idea of making a second relocation. Recently, with the acceleration of urbanization in Ningxia, the land in the near suburbs of cities has been extensively expropriated. Many of the self-initiated migrants have become landless farmers again. They have to face the problem of a second migration.

10.2 Predicaments for Social Governance of Self-initiated Migrant-Inhabited Areas in Ningxia

Self-initiated migration is an inevitable outcome of population mobility following the reform and opening up in China. Self-initiated migration helps relieve the ecological environmental stress in the original places and resolve the migrants' own poverty problems on the one hand. On the other hand, the many predicaments facing the self-development of the self-initiated migrants have also brought a lot of social problems to the host place, as well as great pressure to the social governance at the grassroots government levels. Regarding self-initiated migration, if it were let-go or put-off for a prolonged time, this would be equal to excluding this part of the population outside the scope of the social governance of the government. Nor would they have access to the national treatments they are entitled to. Recently, the number of rights protection-related mass petitions by self-initiated migrants has stayed at a high level in the western regions. Self-initiated migration problems have figured very prominently in the western regions. Self-initiated migration has become a hard nut for the grassroots social governance in the western minority regions.

Prominent problems with the social governance in self-initiated migrant-inhabited areas in Ningxia mainly include the following.

10.2.1 Private Trade of Farmland and Houses, Unauthorized Occupation, and Construction are Heavily Present, Resulting in Disorderly Development and Resources Wastage in Rural Communities

People–land contradiction is the most prominent problem for the host places of the self-initiated migrants. The means by which self-initiated migrants acquire their farmland and homestead in the host places are extremely complicated. Typically, the following means are used. First, the original inhabitants in the host place who have left to run business or work in cities transfer their land or homestead to the self-initiated migrants in private through a compensated means such as sale or lease. Second, some individuals, companies or enterprises, and public institutions reclaim state or collective-owned uncultivated mountains or land at the encouragement or guidance of the national policies and then transfer the land to the self-initiated migrants through a compensated means such as sale or lease. The most special examples are the self-initiated migrant-inhabited areas formed through land development and transferred by individuals having certain economic foundations and social relations, such as the Ximayin Migrant Development Zone in Xixia District and the Xiyanghong Self-initiated Migrant-Inhabited Area in Minning Town, of Yinchuan. Third, the migrants purchase or rent land and houses from enrollees in government-sponsored migration programs, or reclaim uncultivated mountains or land on the periphery of policy migrant-inhabited areas. Some of the self-initiated migrants also conduct secondary sale or lease. Consequently, large areas of state-owned uncultivated mountains and land have been lost; people battle to develop state-owned uncultivated mountains and land, resulting in a disorderly development in the Yellow River-irrigated area of Ningxia. Typical examples include Dahe and Dongyuan Villages of Hongsipu District, Wuzhong City. The self-initiated migrants not only resell homestead and farmland in private, but also build houses without permission. In Laochi Group of Shilipu Village in Xingjing Town, Xixia District of Yinchuan City, there are currently 1000 households. Among them, only 167 are enrollees in government-sponsored migration programs and the remaining 800-plus households are self-initiated migrants. The self-initiated migrants have built their houses in the flood retarding basin, causing the flood control situation to be exceptionally austere. In some host areas, due to lack of long-term planning, houses have been built along roads and canals. The houses so built are of poor quality. This has both affected the overall residing environment of the villages, but also given rise to numerous neighborhood disputes and social contradictions. Of course, there are also some ecological migrants amid the self-initiated migrant population. They both enjoy the resettlement policies for ecological migration and reside constantly in the host area of self-initiated migration, resulting in a high housing vacancy rate in the ecological migrant resettlement area. For instance, in some of the ecological migrant new villages in Luoping County of Wuzhong City, the housing vacancy rate is as high as 30%. In Hongsipu District of Wuzhong City, 1603 households and 6479 residents are included into the relocation under the 12th FYP ecological migration program, accounting for 27% of the total self-initiated migrant population. This people–hukou separation among ecological or self-initiated migrants has brought tremendous trouble to grassroots social governance.

10.2.2 Giving Unplanned Births is Very Common, Challenging the Grassroots Family Planning Management

As the self-initiated migrants mainly come from the southern mountainous areas of Ningxia, they usually refuse to accept the management and service of the family planning department in the host place by excuse of having no residence registration. Extra-policy births are quite common, causing population growth to become too fast. The governments in the places of residence registration, suffering from frequent people mobility and no handhold to implement management service, also find it hard to enforce family planning. The phenomenon of children from unplanned births remaining unregistered is also very prominent. Among some of the self-initiated migrants, their second generation are already old enough to marry and some have even married and given birth to babies, yet they still have no hukou. People–hukou separation or ambiguity has posed a lot of problems to grassroots population management.

10.2.3 Residence Registration Barriers have Become the Greatest Obstacle for the Development of Ecological Migrants, As Well As the Toughest Job for Grassroots Government in Social Governance

The difficulty with the residence registration management of self-initiated migrants primarily lies in the "hukou without people" in the original place and the "people without hukou" in the host place. Some households have hukou in both places; some have more than one hukou; some hold both agricultural residence registration and nonfarm residence registration. As the self-initiated migrants cannot be normally registered in the host place, they cannot enjoy the various benefits provided by residence registration, such as civil reliefs, agricultural subsidies, medical aids, or the various public service and people-benefiting policies such as poverty alleviation programs, irrigation, power supply, road travel, religious venues, and financial credits. Particularly, the self-initiated migrants have to go back to their native place to pay or reimburse endowments and medical insurances. This not only increases the living cost of the migrants, but also raises the probability of their insurance interruption. Of course, there are some self-initiated migrants who choose a "migratory bird" lifestyle. That is, they both enjoy preferential policies provided for returning farmland to forests, returning grazing land to grassland, family planning, and favorable scoring policy in college entrance examination in the original place of residence, but also develop and cultivate land in the present place of residence and, for their own sake, refuse to enter into the local residence registration in the host place. For the grassroots governments, as people and hukou are separated and information is hardly controllable, government management is "vacuumed". The self-initiated migrants refuse to perform their mandatory obligations such as fund or labor raising for infrastructure construction and refuse to accept the management of the local government. Yet, when their own interests are concerned, they would usually contain the government or go on collective petition, heavily affecting the normal work at the village and town levels.

10.2.4 Infrastructure in Self-Initiated Migrant-Inhabited Areas is Extremely Weak, Affecting the Production and Life of the Migrants and Triggering Many Social Contradictions

As little or basically no investment has been made by the government, the self-initiated migrant-inhabited areas usually suffer from inadequate infrastructure, old, out-of-repair roads, and serious lack of public activity sites. Many self-initiated migrants are still drinking canal water or eating cellar water. In some inhabited areas, the irrigation water facilities are unmatched and the canal systems are aged. Some are even transmitting water through earth canals, which are heavily leaky and add to the production cost. In some inhabited areas, disorderly land development and increasing irrigation area have caused the original water right quota to be badly under-provided. Water supply contradiction becomes increasingly prominent. Fighting for water, monopolizing water, and stealing water have frequently taken place. A lot of water-related disputes have arisen out of irrigation problems.

10.2.5 Grassroots Organizations Are Inadequate; Religious and Sectarian Problems Take Place Every Now and then

For many years, governments at various levels hold different understandings and attitudes and lack communication over matters related to self-initiated migration. This has led to severe lack of necessary government duties and severe absence of service management toward self-initiated migration. As a consequence, there have appeared many organization, service, and management "dead zones". The self-initiated migrant-inhabited areas have become a "hot potato" with heaps of contradictions and piles of problems. As the self-initiated migrants mainly come from the southern mountainous areas of Ningxia and the vast majority of them are Islamist Hui people, also because there are a variety of different sects within the same religion, friction is very likely to occur between self-initiated migrants and the host residents who belong to different sects. Furthermore, as grassroots organizations are inadequate, it is usually impossible to eliminate these contradictions in the bud, creating more instabilities for the society.

10.2.6 The Self-Initiated Migrants Are Undereducated with Limited Employment and Entrepreneurship Abilities

The self-initiated migrants are commonly undereducated with limited employment and entrepreneurship abilities and narrow job prospect. They mostly work as physical labor in construction or catering sectors. Sustainable increase of their income is highly difficult. Moreover, weak economic foundation, lack of funds and skills, and insufficient vocational training all add to the difficulty in the employment and entrepreneurship of the migrants.

10.3 Significance of Speeding up the Resolution of Self-initiated Migration Problems

Currently, correctly recognizing and assessing the autonomous relocation behavior of self-initiated migrants is of very great significance to speeding up the resolution of self-initiated migration problems and achieving the targeted poverty eradication objectives.

First of all, relocation of self-initiated migrants has contributed positively to the counter poverty efforts of the western minority regions. In order to resolve their own poverty problems, the self-initiated migrants have relied on themselves and worked hard. After one or even two generations' efforts, the vast majority of them have stepped on the path to stable poverty alleviation and prosperity. In their relocation, the self-initiated migrants not only relieve the tension of people–land contradiction in the original place and contribute to the ecological restoration in the original area but also lighten the many years' of anti-poverty pressure for the government. Compared to the tremendous input made by the state in policy poverty alleviation migration development, their arduous entrepreneurship practice and spirit should be acknowledged. The self-initiated migrations in the western minority regions all have their regularities. Their social benefits should also be evaluated objectively and comprehensively.

Second, leaving self-initiated migration problems unresolved for a prolonged time will result in increasingly serious social problems. Leaving thousands of self-initiated migrants outside government jurisdiction and social governance is not a good thing for either the self-initiated migrants or the society and country. Leaving self-initiated migration problems unresolved for a prolonged time will exacerbate the psychological imbalance of the migrants, giving rise to more social problems in the self-initiated migrant-inhabited areas such as reselling land in private, occupying land, fighting for water, mass petition, giving unplanned births, religious frictions, and illegal and criminal conducts. As the vast majority of the self-initiated migrants are Hui people, this is neither good for guiding religious cultures to adapt to a socialist society nor for the social integration of the self-initiated migrants into the host places.

Finally, the historical opportunity for resolving self-initiated migration problems has become mature. After the third plenary session of the 18th Party Congress, the strategic decision of the central government on comprehensively deepening rural reform explicitly set out the policy of promoting the orderly circulation of the right to the contracted management of land. This laid the political and legal basis for confirming the rights of the homestead and farmland acquired by self-initiated migrants through circulation. The central government has explicitly proposed that by 2020, we should ensure that all the impoverished rural populations will get rid of poverty, ensure that all impoverished counties will shake off poverty, and achieve regional overall poverty eradication. Lifting all impoverished populations out of poverty has become an iconic indicator for building a moderately prosperous society in all respects. By the end of the 13th FYP period, if Ningxia wants to lift all impoverished rural populations out of poverty and win the tough battle of precision-targeting poverty alleviation, self-initiated migration will be a compelling problem that cannot be circumvented. Currently, the party committee and government of Ningxia Hui Autonomous Region are launching policies over the handover of the Ximayin Self-initiated Migrant-Inhabited Zone in Xixia District of Yinchuan City. This move will gain experience for Ningxia to resolve self-initiated migration problems. It will also provide references for the western minority regions to resolve self-initiated migration problems in all respects.

10.4 Paths and Recommendations for Social Governance of Self-initiated Migrant-Inhabited Areas in Ningxia

10.4.1 Guidelines on the Resolution of Self-Initiated Migration Problems in Ningxia

Currently, governments at all levels should, aimed at building a moderately prosperous society in all respects, go all out to resolve the self-initiated migration problems. They should uphold the "respect history, put people first" principle and proceed in conformity with the "digest the stock and control the increment; solve the problem as a whole and implement the policy by classification; grasp the key points and guarantee the people's livelihood" guideline. They should discard the narrow sense of local protectionism of sticking to the national administrative divisions and implement positive social policies.

"Digest the stock and control the increment". The first task is to prevent the unlimited sprawl of self-initiated migrant settlements, set up grassroots organizations in the self-initiated migrant-inhabited areas as soon as possible, and strengthen management service work. For centralized, contiguously inhabited, sizable self-initiated migrant settlements, territorial management should be identified and the settlements should be put under the jurisdiction of the host place.

"Solve the problem as a whole and implement the policy by classification". The self-initiated migration within the province should be considered from all possible angles and resolved as a whole. In terms of the sequential order of resolution, the first should be the sizable self-initiated migrant population residing centrally and contiguously in the host place; the second should be the self-initiated migrants residing in the policy migrant areas with a relatively high share of population; the third should be the sporadically filled-in self-initiated migrants. For self-initiated migrants from outside the province, efforts should be employed to cause the recognition and resolution through coordinated scheming on the national dimension.

"Grasp the key points and guarantee the people's livelihood". Land and residence registration, as the hard nut and focus for resolving self-initiated migration problems, are mutually implicated. Government departments concerned should cooperate closely and fully. They should take ensuring people's well-being as the starting point and goal of their work and earnestly safeguard the legitimate rights and interests of self-initiated migrants.

10.4.2 Recommended Actions to Resolve the Self-initiated Migration Problems in Ningxia

10.4.2.1 Speed Up the Implementation of Residence Registration Verification and Transfer so that the Self-initiated Migrants Have Legal Identity in the Migration Destination

We should resolve the residence registration of self-initiated migrants according to their different conditions, namely whether they are included in the 12th FYP ecological migration program, whether they possess land and house, and whether they come from within the region or outside the region. First, for self-initiated migrants with homestead and contracted land in the host place who voluntarily re-register their residence in the host place, they should first provide supporting materials of the land issued by the land department and agricultural and animal husbandry department of the host place, as well as supporting materials proving their self-initiated waiver of their homestead and contracted land in the native place before they can apply for resident hukou in the host place. Second, for self-initiated migrants with rural homestead in the host place who voluntarily re-register their residence in the host place, they should first provide supporting materials of the land issued by the land department of the host place, as well as supporting materials proving their self-initiated waiver of their homestead in the native place before they can apply for resident hukou in the host place. Third, for self-initiated migrants with contracted land in the host place who voluntarily re-register their residence in the host place, they should first provide supporting materials of the land issued by the agricultural and animal husbandry department of the host place, as well as supporting materials proving their self-initiated waiver of their contracted land in the native place before they can apply for resident hukou in the host place. Fourth, for self-initiated migrants without neither homestead nor contracted land in the host place, the host place may include them into the floating population management and issue them a residence permit according to the relevant provisions concerning residence registration management. Fifth, for self-initiated migrants relocated from outside Ningxia, the host place is recommended to manage them as floating population and issue them a temporary residence permit or residence permit. Sixth, for any self-initiated migrants qualified for the 12th FYP ecological migrant relocation in the native place who has not enjoyed the preferential policies for ecological migrant relocation, in addition to being permitted to have their residence registered, they should also enjoy the 30,000-yuan entrepreneurship subsidy provided by the autonomous region for ecological migrants. Seventh, for self-initiated migrants having enjoyed the ecological migration policies, they can only choose to register their residence in one place. Eighth, for any self-initiated migrants having transferred their residence registration in the host place who are enjoying the returning farmland to forests policy, their original residence registration should be unconditionally written off, the land contract in the original place of residence should be terminated, the homestead use permit should be nullified, and the land should be submitted to the local government for ecological restoration.

When transferring residence registration, we should implement whole-household relocation according to the actual number of members of the family instead of relocating a half and leaving a half in the native place. For those not having had their residence registered due to historical reasons (including marriage, unplanned births, or out-of-wedlock births), their residence should be examined and registered in the native place first and then transferred to the host place. Also, people having their hukou in town should be strictly forbidden from re-registering their residence in rural area. For self-initiated migrants who voluntarily relocate and re-register their hukou in town, their homestead and contracted land should not be withdrawn as a premises for their residence registration in town, as is prohibited by the "Opinions of the State Council on Further Reform of the residence registration System" and the "Implementation Opinions of the Ningxia Hui Autonomous Region People's Government on Further Reform of the residence registration System".

After the residence registration of the self-initiated migrants has been transferred, actions should be taken to speed up the resolution of their livelihood security. First, bring into effect the social security policy. The endowments and medical insurances of the self-initiated migrants should be included into the social security service system of the host place. For destitute households, such as the severely disabled, people receiving basic living guarantee or the five guarantees, orphans, the entitled, and seniors, we should switch their security information to the host area and include them on the jurisdiction's security list of destitute people. Second, bring into effect all people-benefiting policies. All people-benefiting policies concerning employment and low-rent housing should be effected to eliminate the worries and fears of the migrants, and ensure that the migrants can enjoy equal policy treatments as the host residents. For self-initiated migrants engaged in agriculture, subsidies in respect of grain (direct subsidy), fine breeds, agricultural implements, fertilizers, and home appliances going rural should be actively effected. For self-initiated migrants engaged in the second or third industries, tax exemption or tax reduction should be implemented for the first three years. We should strengthen the support of small-sum guaranteed loans from financial institutions for migrants and resolve their shortage of entrepreneurship funds through means such as farmers' joint liability loans and financial discount loans. Third, bring into effect the precision-targeting poverty alleviation policy. Impoverished households among the self-initiated migrants should be precisely identified by reference to the standard for identifying filed and tagged households. Qualified impoverished self-initiated migrant households should be supported by giving them equal policies as for the filed and tagged households to lift them out of poverty and ensure that poverty alleviation leaves no single person out.

10.4.2.2 Effectively Confirm the Rights to Farmland and House/homestead to Practically Protect the Economic Entitlements of the Self-initiated Migrants

Land is the hard nut and core for resolving self-initiated migration problems. For land users not having their residence registration transferred to the host place, no land ownership will be identified. For self-initiated migrants included into the local residence registration management, the ownership of their land, homestead, and house should be confirmed. First, where the actually occupied farmland or house (homestead) is state-owned or collective land in the host county (city, district) with clear, undisputed circulation formalities, the ownership should be confirmed by the government of the host county (city, district). Second, where the actually occupied farmland or house (homestead) is the land of a state-owned farm or wood farm with clear, undisputed circulation formalities, the land should be taken back by the autonomous region government and reallocated to the government of the host place first. Then its ownership should be confirmed by the government of the host place. When taking back and reallocating land, the autonomous region government should give due consideration to the subsequent land demand for public facilities and farmland infrastructural construction in the residing place of the self-initiated migrants and add a proper amount of land to leave room for the planning and development of the self-initiated migrant villages or towns.

10.4.2.3 Resolve Infrastructure Construction Problems to Create A Sound Development Environment

We should include the infrastructure, village planning, industry development, and public service facilities of self-initiated migrant villages and towns into the local socioeconomic development programs, make preferential arrangements, and accelerate their implementation. Currently, the most urgent task is to resolve the construction of infrastructure such as drinking water, power supply, roads, and activity venues for village-level organizations. We should implement rural safe drinking water projects, install safe drinking water facilities, connect to tap water, and resolve drinking water for human and animals. We should enhance rural road construction and comprehensive environmental governance, provide all necessary public activity venues, and gradually improve the human settlement. For migrants residing in crude or badly damaged houses, and those having put up houses without permission before and desiring to build houses, we should guide them to build houses according to the planning, grant them farmhouse renovation subsidies, and better the residing environment and development environment of the self-initiated migrants.

10.4.2.4 Establish and Maintain Grassroots Organizations in Self-initiated Migrant-Inhabited Areas to Provide Handholds for Grassroots Governance

To strengthen and standardize social governance of self-initiated migrant-inhabited areas, it is important to include self-initiated migrants into the current system of administrative control, provide legal guarantee for the rights to autonomy of the self-initiated migrants, and enhance their consciousness and initiative of villager autonomy. Areas where self-initiated migrants are relatively concentrated should be included into local management of the towns (townships) nearby. The host towns (townships) should, according to the conditions for setting up grassroots party organizations, set up a temporary party organization in the migrant area and identify the responsible person to communicate and carry through the party guidelines and policies and maintain the stability of the jurisdiction; where the area does not qualify for setting up a village-level party organization, it should be included into the nearby village committee for management.

10.4.2.5 Bring into Effect the Family Planning Policy According to Law to Control Irrational Population Growth

After the self-initiated migrants have settled down, the family planning policy of the host place should be strictly implemented. Considering that different family planning policies are enacted between the mountainous and river areas of Ningxia, a grace period could be provided. For self-initiated migrants from the southern mountainous areas of Ningxia, within three years after their hukou is transferred, the family planning policy of the native place will still be implemented, management service being provided by the government of the host place. After the three-year period, the family planning policy of the host place should be implemented. Only-child self-initiated migrant families and those qualified under the "fewer children equals faster prosperity" policy should be entitled to priority or preferential policies of family planning "three systems" (reward and support system for eligible families that comply with family planning regulations, the fewer children equals faster prosperity project, and the special support for eligible families that comply with the family planning regulations).

10.4.2.6 Strengthen Industry Development and Support for Entrepreneurship, Try Every Means to Increase the Income of the Migrants

We recommend that the autonomous region should make overall arrangement according to the industry support policies for ecological migrants, provide subsidies to self-initiated migrants engaged in planting or breeding, and include the development of the migrant areas into the poverty alleviation program to effect whole-village advancement. They should intensify educational training and vocational instruction services for self-initiated migrants to enhance their overall qualities and entrepreneurship and employment abilities. They can give the migrants special fund subsidies by reference to the policies for the training of ecological migrants and provide free training on employment skills and practical agricultural skills for them. They should provide free consultation on employment and entrepreneurship policies, employment information, job wanted registration, vocational instructions, and introduction to help the self-initiated migrants to integrate into the local society.

Afterword

The research on ecological migration was jointly conducted by the Institute of Sociology pf Chinese Academy of Social Sciences, Ningxia Academy of Social Sciences, and Beifang University of Nationalities commenced in 2012, when the three institutions made sample surveys and typical case studies on ecological migrants and completed a book "Ecological Migration and Poverty Reduction: A Case Study of Ningxia Hui Autonomous Region in China" based on field investigation. That research was carried out right at the end of the 11th FYP period and the beginning of the 12th FYP period. During the 12th FYP period, Ningxia has experienced the largest-scale ecological migration in history. In order to find out about the impacts of this large-scale population relocation on the host residents, with the support of Bureau of Scientific Research Management of Chinese Academy of Social Sciences, the three institutions have performed a second large-scale survey on ecological migrants in Ningxia, in an attempt to, based on extensive empirical investigations, give an answer to what changes have taken place in the status of the ecological migrants in Ningxia over the past five years, whether the new migrants have adapted to the post-migration socioeconomic environment more quickly with the improvement of migration conditions, and what experiences and inspirations the ecological migration of Ningxia have offered against the general background of precision-targeting poverty alleviation.

The research project was established early in 2016. After that, teachers and students from Beifang University of Nationalities started off a questionnaire survey on 800-plus households of 10 migrant villages. Experts from Institute of Sociology of Chinese Academy of Social Sciences and Ningxia Academy of Social Sciences also launched individual special surveys. At the end of the year, the first draft of the report was completed.

In January 2017, an expert review was organized and the report was highly acclaimed by the experts present. After the review, we supplemented and refined the contents and structure of the report based on the reviewers' opinions.

Our being able to complete the research with high quality in a very short is first of all ascribable to the strong sense of responsibility and sincere cooperation of researchers from individual research institutions, who, with full understanding of

X. Wang (ed.), *Ecological Migration and Targeted Poverty Alleviation in Ningxia*, International Research on Poverty Reduction,
https://doi.org/10.1007/978-981-19-7888-3

the significance of the research, have thrown themselves into the research, pulled together, and contributed their years' academic gains. Second, we feel indebted to the Chinese Academy of Social Sciences for their overwhelming support. Vice President Li Peilin has always been concerned about this research and kindly written the foreword to this report. Vice Director-General of Bureau of Scientific Research Management Wang Zihao has given us careful guidance during the establishment, management, and review of the project. Next, we are gratitude to officials of Poverty Alleviation Office of Ningxia Hui Autonomous Region and local officials at the survey sites for their kind assistance. Finally, we also want to thank the reviewers from Central University of Finance and Economics, Beijing Normal University, and Ministry of Science and Technology for their valuable instructions and useful opinions.

The development of Ningxia Hui Autonomous Region offers an epitome of the development of West China, having gained rich experiences whether in economic development, environmental protection, or poverty alleviation. Our research on Ningxia will not terminate at ecological migration; we also hope to probe into more dimensions of the region over ecological protection, poverty alleviation, and economic development.

SPRINGER NATURE

GPSR Compliance

The European Union's (EU) General Product Safety Regulation (GPSR) is a set of rules that requires consumer products to be safe and our obligations to ensure this.

If you have any concerns about our products, you can contact us on ProductSafety@springernature.com

In case Publisher is established outside the EU, the EU authorized representative is:

Springer Nature Customer Service Center GmbH
Europaplatz 3
69115 Heidelberg, Germany

The manufacturer's authorised representative in the EU is Springer Nature Customer Service Centre GmbH, Europaplatz 3, 69115 Heidelberg, Germany. If you have any concerns regarding our products, please contact ProductSafety@springernature.com

Printed and bound by CPI Group (UK) Ltd, Croydon, CR0 4YY
30/09/2026
02225323-0003